THE MONSTER'S BONES

ALSO BY DAVID K. RANDALL

Black Death at the Golden Gate

The King and Queen of Malibu

Dreamland

THE
MONSTER'S
BONES

*The Discovery of T. Rex and
How It Shook Our World*

DAVID K. RANDALL

W. W. NORTON & COMPANY
Independent Publishers Since 1923

For information about permission to reproduce selections from this book,
write to Permissions, W. W. Norton & Company, Inc.,
500 Fifth Avenue, New York, NY 10110

For information about special discounts for bulk purchases, please contact
W. W. Norton Special Sales at specialsales@wwnorton.com or 800-233-4830

Manufacturing by Lakeside Book Company
Book design by Lovedog Studio
Production manager: Lauren Abbate

ISBN 978-1-324-00653-4

W. W. Norton & Company, Inc., 500 Fifth Avenue, New York, N.Y. 10110
www.wwnorton.com

W. W. Norton & Company Ltd., 15 Carlisle Street, London W1D 3BS

1 2 3 4 5 6 7 8 9 0

To Diane Randall,
who took me to every museum she could

Nature has a habit of placing some of her most attractive treasures in places where it is difficult to locate and obtain them.

—CHARLES DOOLITTLE WALCOTT,
FOURTH SECRETARY OF
THE SMITHSONIAN INSTITUTION

CONTENTS

THE CENTER OF THE WORLD

DEPENDING ON WHICH ENTRANCE YOU CHOOSE, THE American Museum of Natural History looks like a castle the color of dried strawberries, a sun-bleached Roman temple or a spaceship ready to launch out of a glass box. The fact that its sides don't match is the first clue that, like the millions of once-living specimens that can be found within its maze of twenty-eight interconnected buildings in the middle of Manhattan, the museum grew organically, with something close to chance as a guide. Since its founding in 1869, the museum has gathered, categorized and displayed all that is amazing, strange and beautiful about the planet Earth and the universe in which it spins. To tour its collection is to viscerally experience the sum of human knowledge of the world surrounding us and our best guesses at how we fit into the picture.

The numbers alone are so large as to seem made up. Among the museum's possessions are nearly seven million wasps, three million butterflies, one million birds and over two million fish, though they are no longer swimming. And that's just a modest survey of its extensive library of small animals. If the contents of the museum were taken across the street and laid out in Central Park like a gigantic yard sale, you could have your pick from the bones of 100 elephants, including the skeleton of Jumbo, once the most famous zoo animal in the world and the reason why the English language gained a new word for something particularly big; one thirty-four-ton meteorite called Ahnighito that is nearly as old as the Sun; more than 1,000 representations of the Buddha; 45 musical instruments from the Congo; the stump of a 1,400-year-old sequoia; 51 balls used in sports ranging from lacrosse to a game played by the Tupari people of South America, in which players could only touch the ball with their heads; the skeleton of a 3.1-million-year-old hominid woman that scientists named Lucy after a Beatles song that was playing on the radio; a set of dinosaur footprints that are at least 107 million years old; a 563-carat blue star sapphire known as the Star of India; a taxidermied Galapagos tortoise; a shirt woven by the Cheyenne decorated with locks of human hair; and a 4.5-ton block of azurite–malachite ore from Arizona that looks as if Jackson Pollock decided to paint a rock bright blue and green and which contains so much copper that it can vibrate and hum when the air becomes very humid, giving it the nickname the Singing Stone. Not to mention the dinosaur eggs, the world's largest collection of spiders, the bones of the biggest hippo ever found, a 12-foot-long crab, a 55-pound slab of jade, a 63-foot-long canoe carved from the trunk of a single cedar tree and a life-size model of a blue whale hanging from the ceiling, which has its mouth closed because the chair of the Mammalogy Department wanted to prevent visitors from trying to throw things inside it.

How each wonder of the natural world came to sit in a museum in one of the most densely populated places on Earth is its own story. Yet even with an array of the splendid and surreal, it is no exaggeration to say that only a slice of the more than five million people who come to visit the collection of the American Museum each year would bother at all if it were not for one specimen: the *Tyrannosaurus rex*. It is the most iconic dinosaur in the world, the one creature for which seemingly everyone knows its full Latin name. Though the last of its kind died more than 66 million years

ago, the massive jaws, tiny forearms and long teeth of a *T. rex* are
as instantly recognizable as the Mona Lisa or the Eiffel Tower,
its deadly beauty coming from a seamless pairing of form and
purpose. It is not every day that one confronts a huge beast that,
should it somehow become reanimated with muscle and memory,
would clearly like nothing more than to crush your bones with its
serrated teeth and eat you up in a gulp or two, like the monster
from a child's nightmare made real.

For thirty years, the museum was the only place in the world
where you could view a *T. rex*, and the specimen has called Manhattan home since before the construction of the Empire State
Building. To a degree of no other object that can be found in a
museum, the *T. rex* is quilted into modern culture, a fixture of
everything from clothing to toy stores to blockbuster movies. Its
familiarity makes it seem like it has always been with us. Yet there
was a time not very long ago when everything about this prehistoric beast was entirely new, and the thud with which its discovery
landed and shifted our understanding of ourselves and our planet
reverberates still.

It seems odd to suggest that a species that last lived at least 60
million years before the appearance of the first species that could
be called human ancestors somehow changed the course of modern life. The appeal of the *T. rex*, after all, rests in part on the
fact that it is long dead, of no more mortal concern to us than
a scary movie. It took a family trip to the American Museum
one brisk fall morning for me to consider how a massive dinosaur
once buried under layers of hard rock in Montana ended up in the
middle of Manhattan. Growing up, I somehow dodged a dinosaur phase, never becoming the sort of person who could discuss
with confidence why despite all appearances to the contrary the
sail-backed prehistoric creature *Dimetrodon* was actually a distant
relative of primitive mammals and not a dinosaur at all. With
our kindergartener and preschooler in tow, my wife and I spent

more than an hour walking through the dinosaur halls, though we somehow continued to loop back to the *T. rex* again and again. On the last lap, our son grew quiet, leaving us to exchange glances: perhaps the specimen, which had teeth nearly as long as his torso, was a little too frightening after all. Then he turned to us and asked, "Who found these dinosaurs?"

It was the first time I considered the human stories behind prehistoric bones. It soon became clear that in no small way the discovery and eventual display of the *T. rex* not only saved the American Museum from becoming an afterthought to the competing Museum of Modern Art on the other side of Central Park, but played a large part in bringing an appreciation and understanding of natural science into mainstream life. Dinosaur bones, of course, had long been a part of the natural landscape. But in the span of a generation, the work of a small group of men and women trying to find meaning in the remnants of a former world turned fossils from a biological curiosity into an attraction that brought millions into newly-opened museums to learn more about the mind-bogglingly long history of life on Earth. At a time when the notion of an unrecognizable Earth dominated by dinosaurs was a revolutionary idea, the *T. rex* resonated with the public like no other creature before it, the star whose name is listed at the top of the marquee.

This book tells the story of the discovery of one of the largest known predators in Earth's history and the ways in which it continues to affect our lives. It is the tale of a spectacular competition among pioneering scientists, opportunistic cowboys and some of the wealthiest men the world has ever seen to find and display the largest dinosaurs on record, pushing our understanding of the planet and our place on it to new heights. The *T. rex*, the king of the prehistoric world, played an outsized role in the making of our modern one: robber barons saw in its carnivorous ferocity a justification of their own elevated powers; eugenicists found evidence of prehistoric domination which they used to support noxious theories of racial superiority; and ambitious dinosaur hunters discovered a path out of the narrow confines of their lives. Rather than a mirror into the past, the creature proved to reflect the concerns of the present.

Through their well-organized floors and colorful displays, museums give the mistaken impression that the history of life follows a narrative that is mannered and coherent, moving forward gracefully to an obvious endpoint. In reality, it is something closer to chaos, a series of random events that only in retrospect appear to have a destination in mind. The story of *T. rex*—a tale that spans from the highest levels of wealth in the twentieth century to the poorest shacks on the frontier—is the story of how we came to understand our planet and were offered the first glimpses into the power of a changing climate to eradicate the dominant forms of life.

And at its center was a boy from the open plains of Kansas who stood with one foot in the frontier and one in the modern world, seeing in the deep past a way to ease his pain and remake his future.

THE MONSTER'S BONES

Chapter One

A Life That Could Contain Him

THE BOY NEEDED A NAME.

It was a duty that William and Clara Brown had already answered three times before, following the births of their two daughters and a son. Now, with another infant staring up at them in the farmhouse where they lived on the outskirts of Carbondale, Kansas, it seemed an overwhelming task. A name was supposed to guide the child on a safe and righteous path, instilling trust in others and resolve in oneself. William, especially, was highly attuned to any sign that someone was deviating from a Christian way of life, often refusing to sell coal to buyers in distant cities after deciding that their penmanship reflected poorly on their character. And besides, all of their good ideas had been taken. Their two daughters had been named after family, closing off that avenue of inspiration. At the age of six, their first son, Frank, was as forthright as his name suggested, developing into such a miniature version of his father that the pair were clad in matching three-piece suits and poorly-tied bow ties in family portraits. Yet here before them lay another boy, his fate so uncertain that several days after his birth he still didn't have anything to shape his identity.

Leaving so weighty a decision to the last minute was entirely out of the Brown character. Like all farmers, William prized certainty where he could find it, seeking any comfort from the annual cycle of planting crops in the spring without knowing whether there would be enough summer rain to nurture them. Born in Virginia in 1833, he migrated to the open plains of the West as a twenty-one-year-old in search of a better chance. "There were those among the pioneers who were merely drifters, fiddle-footed and restless, that wandered westward either to escape an unpleasant situation in the east, or in the hope of getting something for nothing in the west. . . . Father's pioneering was purposeful: he was hard-working, with a good head for business; he sought and found promising opportunities worthy of the heavy investment of thought, time, and labor that he poured into them," his youngest son later wrote. In Wisconsin, he met Clara Silver, the fifteen-year-old daughter of a prosperous dairy farmer, and the two were married by the end of the year. After four years of accumulating livestock, the pair—accompanied by a young daughter and with another on the way—headed toward the Kansas Territory in 1859. They eventually stopped in Osage County, in the eastern part of the state near the Missouri border about twenty miles south of the state capital, where William purchased land on a spot named Carbon Hill, due to the long bands of coal which blackened the soil.

The young family lived out of their wagon while William built them a home. The small wooden house soon turned into a refuge, as the Kansas prairie devolved into a remote battlefield in the Civil War. Pro-Union guerrilla fighters, known as jayhawkers, battled Confederate sympathizers known as bushwhackers, each side ambushing the other in bloody raids that terrorized the countryside. "Mother used to say that during the War it was not uncommon for Federals to stop for food in the morning, Rebels at noon, and Bushwhackers at night. She never dared to say where their sympathy lay for fear of retaliation by shooting the family, burning down their house, or destroying their property," her youngest son later wrote. Clara was able to maintain a neutrality in part because William was away for long periods of time, having secured a lucrative contract to lead wagon-trains full of supplies for the U.S. Army across the open frontier. Once the war was over, he watched as ribbons of railroad tracks appeared on the plains, stitching together the Transcontinental Railroad and boxing in what had seemed to be the limitless horizon.

The rails made him feel like a relic in his own time. Determined not to be left further behind, he hitched up his team of oxen and began clearing off the topsoil of his land, revealing long bands of coal that could be extracted from the earth and sold to buyers across the state. Now in the mining business, William built a modern two-story house with a gabled roof that was soon recognized as "the best residence in this section of the country" by a local committee. Though never rich by the standards of a city, the Browns grew prosperous enough that they employed thirty-one men to help run the farm and dig up coal, feeding everyone three times a day in the family's living room. With any money that was left over, they traveled up to the capital city of Topeka and its diversions.

It was there that six-year-old Frank saw an advertisement for P. T. Barnum's Great Traveling World's Fair plastered on barns, trees, and the sides of seemingly every office building, the bright and bold colors of the circus beaconing through the low winter light. Soon he could think of nothing else. The birth of his younger brother a month later on February 12, 1873, did little to cool the heat of his focus. As William and Clarissa sat at the kitchen table discussing possible names, Frank burst in and yelled, "Let's call him Barnum!" Though it had nothing to offer in the way of family lineage and carried with it a suggestion of showmanship that seemed out of place in Kansas, the name somehow stuck. After several more days in which his parents tried and failed to find a better alternative, a newly-christened Barnum Brown was finally ready to face the future.

✝ ✝ ✝ ✝

THE FIRST YEARS OF HIS life offered few chances to live up to his name. His earliest memory was of lying in a bassinet under an oak tree on a sunny day, spending an unhurried afternoon with his mother listening to the wind rustle the leaves. As soon as he could walk, he was cast in the daily choreography of life on a farm that made the place function. He hauled water, milked cows and helped his mother raise lettuce and onions in the garden outside the kitchen window—nothing, in short, that would suggest the extraordinary life that was to come. His mother began to rely on him so much for help around the house that Barnum would later write that she said that "I was the best 'girl' she ever had, for Melissa was romancing and Alice, who resented being a girl and loved men's work, was raising cattle on her own, helped by brother Frank . . . so little brother got a lot of housework." When he turned ten, he graduated to chores outside of the home. "In the late summer, I sometimes milked 20 cows, morning and night, sitting on a one-legged stool, so if a cow kicked I would fall over without resistance," he later wrote. For company he had a Newfoundland dog, Old Bruno, who weighed twice as much as he did and would often disappear for hours chasing jackrabbits.

When Barnum grew old enough to accompany his father, his favorite pastime was following along behind the team of oxen as they stripped layers of soil off the land in William's never-ending search for new seams of coal. Sometimes as much as eighteen feet of dirt and rock needed to be removed before he found one, leaving makeshift mountain ranges of discarded rubble for Barnum to play on. When the boy wasn't climbing, he was digging, and soon began to notice that some of the objects that at first glance appeared like any other rock seemed instead like they belonged on a beach. One looked like a piece of honeycomb coated in beeswax; another was three inches long and shaped like a perfect cornucopia.

He began holding on to as many of them as he could find, with the help of his father who "though untrained in geology, encouraged me in making these collections, for he thought that by doing so we could find out why sea shells could be entombed in a Kansas hilltop 650 miles from the nearest seacoast today, the Gulf of Mexico," Barnum later wrote. Soon, he was storing so many specimens inside his room that his clothes could no longer fit in his drawers, leading his mother to banish his shells to a nearby shed, where Barnum set up his own showroom—the first but not the last time that he would inadvertently follow in the footsteps of his namesake. "This became my first museum, where I had my first experience as a showman regaling visitors with these treasures, together with the Indian arrow points and scrapers I picked up while plowing our cornfields," he later wrote.

The boy had somehow stumbled onto the most pressing question in science. A hundred years before his birth, the history of the planet barely registered as a serious avenue of intellectual inquiry. In the late 1600s, James Usher, the archbishop of Armagh, had seemed to settle the question by adding together the lifespans of every named descendant of Adam in the Bible and cross-referencing them with the seasonality of the Hebrew calendar, allowing him to conclude that God created the Earth on the night of October 23, exactly 4,004 years before the birth of Christ. This was considered a landmark feat of scholarship, prompting some printers to add dates in red ink on the margins of pages of the Old Testament to give readers an exact chronological grounding. Others sought to emulate Usher's achievement by turning their attention to questions such as how exactly Noah organized all of the animals in the Ark.

Yet the realization that certain kinds of rocks and minerals had outsized economic purpose that could feed the new machines of industry coming out of workshops across Europe sent miners deeper and deeper into the Earth, and they kept coming back with things that shouldn't have been there if God had created the world in an orderly seven days some six thousand years ago, as the Bible suggested. Why were the bones of what appeared to be ancient fish uncovered far below the ground of open plains? Why were strange skeletons that didn't seem to match any living things found poking out of cliffs and marshes? And, if seashells found at the top of high mountains were evidence of the Great Flood, as some scholars argued, then why weren't the surrounding rocks smoothed by the erosion of such an immense amount of water? Rather than the neat process of Creation described in the Bible, Earth seemed to be covered with the scars of chaos.

Miners were among the first to realize that the planet was much older than a literal interpretation of the Bible would suggest. In the middle of the eighteenth century, a young German by the name of Abraham Werner grew up accompanying his father on his inspections of the Duke of Solm's ironworks, and eventually enrolled at the prestigious Freiburg School of Mining. While most of his fellow students were immersed in questions of where to find silver and how to square the mineral formations they found with what they learned in the Bible, Werner noticed that Earth's crust seemed to be made up of four categories of rock that always appeared in the same order.

The oldest and deepest layer he called primary rocks, which contained formations such as granite that contained no trace of life. Above those he identified transition rocks, which included slate and graywacke and the first appearance of fossils. In Werner's time, fossils were widely recognized as something more than strange stones, though what exactly they were remained unclear. Aristotle was among the first to put forth the theory that some material that appeared to be rock was once animated with life. This conception was not universally accepted, however, and centuries later some European theologians proposed that perhaps stones that took the form of bizarre animals or plants was God's way of ornamenting the interior of the Earth, much like flowers decorate its surface, or perhaps they were red herrings placed there by God as tests of faith. In any case, the fossilized relics of life—bones, shells, teeth or leaves—are not in the strictest sense the biological remains of what was once a plant or animal; instead, they are inorganic minerals like calcium that linger after tissues and blood cells have decayed, essentially becoming rocks that are in the exact shape of the original living material.

Werner, however, was concerned not with the history of life on Earth, but with the composition of the planet that sustained it. He noticed that on top of the transition rocks lay what he called the secondary layer, which was made up of sandstone and gypsum and other rock formations that were bursting with evidence of long-dead plants and animals. Finally came the tertiary layer, which consisted of sand and clay and an abundance of subterranean life-forms that could be found by digging in the ground with your bare hands. After studying those repeated, stratified layers of material beneath our feet, Werner concluded that the entire planet had once been a vast ocean from which Earth's crust slowly emerged over more than a million years.

In Scotland, a farmer and naturalist by the name of James Hutton agreed with Werner that Earth had a history longer than anyone thought. So much longer, in fact, that he would later write "that we find no vestige of a beginning and no prospect of an end." Hutton, who had studied chemistry at the University of Edinburgh before running two small family farms, presented a paper in 1788 in which he theorized that the planet's surface was part of a continuous cycle in which sediments on land are carried to the ocean, where they are compacted into bedrock on the sea floor over time until they are ejected once more back above the surface by volcanoes and other eruptions from deep within the earth, starting the process fresh again. As evidence, he pointed to a promontory at Siccar Point on the east coast of Scotland where horizontal layers of red sandstone intermingle with a string of vertical layers of gray shale, an arrangement that, far from Werner's idea that all types of stone are found in a uniform sequence, could only be explained by the violent churn of the planet weaving bands of rock formed on land with those formed undersea over a time period of millions of years.

He described a process in which what appear as unmoving mountains to our eyes were once rocks buried on the bottom of the ocean, a cycle ticking along at a scale of destruction slower and more momentous than we can imagine. "We have the satisfaction to find, that in nature there is wisdom, system and consistency," Hutton wrote. "For having, in the natural history of this earth, seen a succession of worlds, we may from this conclude that, there is a system in nature; in like manner as, from seeing revolutions of the planets, it is concluded, that there is a system by which they are intended to continue those revolutions. But if the succession of worlds is established in the system of nature, it is vain to look for anything higher in the origin of the earth."

Geology, it seemed, was anything but the neat and tidy study of rocks. Instead, it was a window into an abyss of time so deep that it was nearly incomprehensible. Was Earth a million years old? A billion? More? And if the planet was so old and the record of human life in the Bible dated back only six thousand years, what if anything took place during those missing millennia, a gap of the unknown stretching so wide that it may as well have been infinite? Recognizing a threat to their explanation of the world, religious scholars downplayed both the significance and relevance of the dangerous new science. When a newly established professor of mineralogy at Oxford briefly left the university to undertake field studies in Europe, one dean happily announced, "Thank God we will hear no more of this Geology!"

For all of the existential questions it raised about life on Earth and humankind's place within it, geology as a discipline was too valuable to ignore. Hidden inside the Earth were minerals, precious metals and soon-to-be-important oil fields that were worth a lot of money, and treasure hunters cared little about what it all meant in a cosmic sense as long as this new branch of science could be of some use in their goal of getting rich. A year before the United Kingdom undertook a similar project, the United States Congress directed the Topographical Bureau of the U.S. Army Corps of Engineers to construct the first geological map of the young and growing country, in 1834. "Few subjects connected with the duties of this bureau open so many and so important national advantages, or are adapted to redound more to international commercial prosperity [as the] development of these great resources of wealth and commercial interests, which now lie inert and buried in the bowels of the earth," Colonel J. J. Albert, the head of the bureau, said when requesting funding for the project. In 1859, for the first time, the value of industrial products manufactured or built in the country exceeded the value of everything grown or produced on its farms, hastening the search for the raw materials that the new economy relied on.

AS BARNUM BROWN GREW OLDER, a world built on the products of mining erased the slower-paced existence that his parents had known. Iron machines began to replace oxen in the fields; telegraphs stood in the place of handwritten requests for another shipment of coal. The pace of change seemed to sweep him up along with it, instilling in him a sense of restlessness that could never be fulfilled on a farm. It was a feeling that was infectious, a rush of excitement that seemed to emanate off him and enchant everyone he met. In time, his spell would prompt a New York socialite to give up her charmed existence and go with him on a search by foot through rural India for victims of an outbreak of bubonic plague. As a teenager, he was tall, lean, curious and fully aware that his twinkling blue eyes made him handsome—a potent combination incompatible with a town whose main drag consisted of a bank, two churches and a grocery store where they gave away unsold meat for dog food. To be in his company meant unshackling whatever tied you to the present and finding yourself compelled to explore beyond not only the distant horizon but whatever came after that, pushing past everything known until you reached a blank spot on the map. Carbondale, where the greatest excitement came in the form of updated grain prices or the inevitable side effects of a hive of eighteen saloons catering to the town's two thousand people, could simply not keep up with him. There wasn't even a high school to keep his mind distracted from wondering what life was like at the far ends of the train tracks that passed through town.

Barnum's parents knew that any energy spent trying to tether him to the farm was wasted. Once he finished the last year of schooling available to him in Carbondale, in 1889, they made plans for him to move to the university town of Lawrence thirty-five miles away, where he could attend high school before enrolling in the University of Kansas. Before the sixteen-year-old left home, however, his father wanted him to accompany him on one last adventure, a journey whose ostensible purpose was to find land suitable for a cattle ranch but was really an excuse to get out and see the world together while they could. "Father wanted me to see what was left of the Old West before it faded away, to show me some of the places he had been in his pioneer days, and to broaden the outlook of an adolescent farm boy who had never been more than twenty miles from home," Barnum later wrote.

Together, they prepared a covered wagon and packed it with enough sugar, bacon, flour, beans, raisins and coffee to last them four months on the plains. They headed north, waking up each morning before sunrise so that Barnum could prepare them breakfast while his father fed the team of horses, and stopping in the afternoon when they found good grazing land. They would repeat the same process at night, with his father taking the extra step of padlocking the animals to the wagon when he was worried that the smell of Barnum's cooking would bring unwanted attention as they passed through the quilt of lands belonging to the Comanche, Kiowa, Sioux and other tribes who still roamed the Plains in diminished numbers. As they rode, William told his son tales of a world that no longer existed. Bison were once so plentiful that his father had witnessed "great herds . . . streaming across the Missouri River in such numbers as to stop the river boats," Barnum later wrote. The introduction of barbed wire, railroads and bloodthirsty settlers carrying powerful rifles had exterminated millions of bison in the brief span of Barnum's life, leaving the boy to relent that "we saw only their heads, preserved by the long shaggy hair and tough hide, strewn across the prairies like small barrels."

In southeastern Montana, the Browns neared the site of the Battle of the Little Bighorn, where thousands of Cheyenne and Sioux warriors had decimated George Armstrong Custer's force of 250 calvary some thirteen years earlier. As they approached the battlefield on the Fourth of July, a Crow guide spotted them and directed them to a prime viewing spot on the ford of the Bighorn River, where they watched a Native American reenactment of the battle while a nearby garrison of soldiers looked on with their guns ready. They then continued westward across the wide-open land, through what would eventually become Yellowstone National Park. There, Barnum discovered that the lakes were teeming with trout, so many that he could snare two fish with one cast and soon filled up a keg with them. He traded his catch to the men of the U.S. Cavalry, who were not allowed to go hunting or fishing in the hostile country, and came back with a big jar of pickles that he ate in one sitting.

Only after they spotted the snowcapped peaks of the Rockies did William decide that it was time to head back, leaving behind the open country of his youth and the company of his youngest son to return to a present where he felt he had no future. "There were more new sights and every day was a fascinating adventure," Barnum later wrote. "And then we were home again with our loved ones. We had trailed about three thousand miles at an average rate of twenty-five miles a day when we were on the march. It took us a little more than four months from start to finish. What an experience! This was Father's finest gift to me; it was of himself."

In the days after his return, Barnum readied himself for the move to Lawrence and its university, a place that seemed unbelievably cosmopolitan after the long, empty hours on the farm without the company of anyone his own age. As he boxed up his childhood, he packed shells from his personal museum to take with him, intent on finding a professor who could explain why the world suggested by the relics he found underground did not match what he saw in the present day. He did a last round of chores on the farm, hoping that he would never return to an existence circumscribed by duty and the seasons. He was young, ambitious and, thanks to the journey into the wilderness with his father, had the self-confidence to survive in any situation. At the age of sixteen, he set out in search of a life that could contain him.

Chapter Two

A WORLD PREVIOUS
TO OURS

LAWRENCE, KANSAS, IS LOCATED ABOUT TWO HUNDRED
miles east of the geographic center of the continental United
States, and, perhaps due to this proximity, has a long history of
being in the middle of things. The town was settled on the urging
of a congressman from Massachusetts named Eli Thayer after the
passage of the Kansas–Nebraska Act in 1854, which allowed states
newly admitted to the Union to choose for themselves whether
to allow slavery. "Let us settle Kansas with people who will make
it free by their own voice and vote," he argued. "Come on, then,
gentlemen of the slave states. Since there is no escaping your challenge, we will accept it in the name of freedom. We will engage in
competition for the virgin soil of Kansas, and God give the victory
to the side which is stronger in numbers, as it is right." A party of
New England abolitionists traveled along what was then known
as the California Road, a path along which thousands of Gold
Rushers had flowed west a few years earlier, and selected a spot in
the rolling prairie between the Kansas and Wakarusa rivers for a
settlement they named Lawrence.

The fighting began not long after the town had a name. An
abolitionist settler from Ohio by the name of Charles Dow was
shot in the back by his pro-slavery neighbor, Franklin Coleman,

the culmination of an argument between the two men that had lasted several weeks over the boundaries of their land. Coleman fled to neighboring Missouri, where he convinced a sympathetic sheriff to arrest one of Dow's friends who had vowed to avenge his death. That, in turn, provoked a group of at least fifteen abolitionists in Lawrence to form a raiding party and attempt a rescue. "They were armed with all sorts of weapons," wrote Richard Cordley, an abolitionist minister who eventually received the first degree awarded by the University of Kansas. "Some of them had rifles; some of them had shotguns; and some of them had pistols. They had come with anything they happened to have in the house. One or two had no weapons whatever. One of these picked up two large stones which he clutched in his hands in a way which showed his intensity of purpose, and illustrated the determination of the whole company."

Kansas was admitted to the Union as a free state in 1861, three months before the start of the Civil War, though that did little to settle Lawrence's future. Two years later, more than four hundred Confederate guerrillas under the command of a former Missouri schoolteacher named William Quantrill torched the city. It rebuilt itself in months, a flurry of activity that Cordley called "a matter of conscience." Lawrence pressed forward. In 1865, a newly-established Board of Regents officially founded the University of Kansas on a small hill overlooking Lawrence; in 1867, the first locomotive reached the city. "Kansans had suffered from practically every affliction known to man: from violence, murder and territorial civil war, from a bloody and costly national civil war, from drought and famine and disease and poverty, from senseless and unscrupulous political conniving complete with lies, deceit and whole knavery," wrote one historian of the time. "The creation of the university was not exactly a miracle—miracles were rare in Kansas—but it was still an awesome event."

Nearly everything about the existence of the University of Kansas was unlikely, and it reciprocated in kind by filling its halls with the sort of students who were not found in the elite institutions of the East Coast. Across the country, the rapid pace of industrialization meant that the raw material of America was for the first time tapped for its potential at a grand scale, refining the base elements of nature into something greater than was there before. Iron ore became steel; coal transformed into power for steam engines and factories; seams of gold glittered into jewelry. And, at the dozens of public colleges and universities sprouting up across the country, young adults whose parents were never given an opportunity for a formal education became a rising class of professionals ready to build new lives that did not rely on their physical labor. The first student body of the University of Kansas consisted of twenty-six women and twenty-nine men, making it one of the first public institutions in the country to admit women on an equal footing, and the university made it its mission to show that the brains of the children of farmers were just as potent as those of the privileged. To do so, it hired professors who were tired of the politics and traditions of liberal arts institutions on the East Coast and welcomed the chance to do things differently, often inserting Lawrence into intellectual disputes with universities founded before the Revolutionary War.

It was a place ideally suited for a young man in a hurry like Barnum Brown. He entered the university as an engineering student in the fall of 1893 after completing two years of high school in Lawrence, setting up his fossil collection in his shared room on the second floor of the only dorm on campus. After years of isolation on the farm, the fact that he was now surrounded by peers was intoxicating. It would soon become clear that despite his intellect Brown was not a natural scholar, finding himself more attuned to the social aspects of the university than to its academic demands. He would tell stories about his college days for the rest of his life, often recounting his adolescent adventures with such joy that it seemed as if they had happened yesterday.

In notes for his never published autobiography, he described how a friend named Baker had forgotten the terms "oviparous" and "viviparous" in zoology class, leading him to stand up and ask the professor, "Professor Dyche, do bats lay eggs or do they ah-ah?" When the professor replied, "Mr. Baker, the bats ah-ah," Brown wrote that "the girls figuratively swallowed their handkerchiefs and Baker had to leave school for a week to escape ridicule." In chemistry, Brown learned that metallic sodium would ignite if exposed to water. He put this knowledge to work by sneaking with his roommate into the dorm room of the women who lived below them and laying a coil of the substance in their chamberpot while they were away. In the middle of the night, "a scream came out of their room for one the girls thought she was sitting over a sizzling snake."

Amid the juvenile pranks, Brown found time to pursue the explanations that had eluded him as a boy. The University of Kansas was still new, and that sense of being at the beginning encouraged students to seek out those who could answer their questions. A box full of seashells he'd found at his family's farm sat in his room, a daily reminder of what had once animated him to stretch beyond the limits of home. Though he had yet to have him as a teacher, Brown made his way to the office of the professor whom he would credit with changing his life.

Like Brown, Samuel Wendell Williston was the son of migrants who came to Kansas in search of better opportunities, and he spent his early life crammed in a one-room cabin with a family of six who never had enough to eat. Despite the hardships of the frontier, Williston was naturally more bookish than his brothers, and at the age of fifteen he escaped to the Kansas State Agricultural College, where he began collecting insects and fossils. Thanks to the recommendation of Benjamin Franklin Mudge, a renowned professor at the university who wrote the first book on the geology of Kansas, Williston was hired as the primary assistant to Othniel Charles Marsh at Yale. There, he had a front-row seat to a bitter feud that greatly expanded the number of prehistoric creatures known to have once lived on the Earth, conducted at the cost of one man's fortune and the other's reputation.

THE TERM "DINOSAUR" DATES TO April 1842, when a wealthy professor of anatomy named Richard Owen published a paper in which he argued that a new term was needed for the strange, enormous bones that were popping up throughout England. He landed on it by combining the Greek word *deinos*, which not only means terrible or fearfully great but also implies a sense of the inconceivable, with *sauros*, which means lizard. The animals that made up the order of *Dinosauria* "attained the greatest bulk, and must have played the most conspicuous parts . . . as devourers of animals and vegetables, that this earth has ever witnessed," he wrote. And then, in a moment of self-congratulation altogether in character for a tall, intense man known for his unblinking stare, he added, "A too cautious observer would, perhaps have shrunk from such speculations. . . . But the sincere and ardent searcher after truth, in exploring the dark regions of the past, must feel himself bound to speak of whatever a ray from the intellectual torch may reach, even though the features of that object should be but dimly revealed."

Owen's decision to label what we now think of as dinosaurs as a new order was inspired less by the intellectual torch than by resentment of a less-connected rival. He grew up in the town of Lancaster as the wealthy son of a draper who had made a fortune in trade with the West Indies, and seemed uninterested in any particular profession, bouncing through life as if intent to make good on his childhood tutor's warnings that he was "lazy and impudent" and would "come to a bad end." With his father's help, Owen secured a position as an apprentice to a local surgeon, and began treating prisoners at a county jail with leeches and other natural remedies. He was soon called to assist in his first autopsy, conducted on the remains of an inmate who died while in the state's care and had no relatives to object to a procedure that was by many considered a sin. Convinced that he had seen a ghost while walking to the jail tower, Owen was late to recover the body, and as soon as the procedure started regretted every decision that had led him to that moment. He felt such an overwhelming sense of revulsion the first time he saw the organs and bones of the dead man whom he had failed to heal that he vowed to "never, never again . . . desecrate the Christian corpse and to quit a profession that could only be learned by such practices so repugnant to the best feelings of one's nature."

Yet something made him stay. He grew fascinated with anatomy, seeing in the perfect architecture of muscle and bone the hand of God. He began collecting the skulls of dogs and cats, as well as any human remains that he could secure through whatever means necessary. (He once paid a jailer for the severed head of a Black prisoner who had died in custody and, while carrying it in a bag back to his laboratory, slipped on some ice; the head went tumbling down a snowy hill and into the open door of a cottage, where Owen raced inside and grabbed it without stopping to calm the terrified people who lived inside. The red spots in the snow the next day led to whispers that the Devil was raising an army conscripted of the ghosts of the enslaved.)

The focus on the anatomical structure of life led Owen to consider the intent of its Creator. Were all species just as God had originally made them, with the strong jaws and narrow vision of a carnivore a sign that it was designed for predation while the hooves and flat teeth of herbivores signaled they were predestined for evasion, or did lifeforms change over time? Already, science had abundant evidence in the form of folklore that the history of life on Earth was not a straight line, though few who considered themselves scholars would stoop to believing that stories of fancy could have a basis in fact.

Ancient discoveries of fossils likely served as the basis for legendary creatures such as dragons and sea serpents, as cultures around the world looked to square the discovery of bones of no-longer living creatures with the fact that they existed at all. Tattoos of griffins—mythical eagle-beaked lions adorned with powerful wings—decorated the skin of nomadic Saka–Scythnians who lived five centuries before the common era and whose mummified remains were discovered in the Altai Mountains of Central Asia in the 1940s. The same sandstone region is now known as an abundant source of *Protoceratops*, beaked dinosaurs the size of sheep whose profiles look remarkably like those of griffins. The fossil beds on the Greek island of Samos, meanwhile, likely inspired tales of savage monsters known as Neades, thought to be able to tear the earth apart with their cries. In a depiction of Heracles rescuing Hesione from the Monster of Troy on a Corinthian vase dating to the sixth century BCE, the monster's skull closely matches that of an extinct giraffe whose remains are widespread throughout Greece and Turkey. In North America, Algonquins recognized what they called "the bones found under the Earth" as ancient demons killed by the heroic god Manabozho, while the Blackfeet considered dinosaur fossils "the grandfathers of the buffalo" and the Zuni saw fossils of belemnites—a form of prehistoric squid—as "lightning or thunder stones" and prized them as a form of battle armor. In Asia, what we now call dinosaur fossils were once considered "dragon's bones" and often ground into fine powders for tonics and medicines, while artwork found in the Mokhali Cave in the southern African nation of Lesotho shows what appears to be a dinosaur leaving footprints across the valley.

Over time, the abundant remnants of prehistoric life became harder to reconcile with the idea that they were the relics of

monsters abolished by ancient heroes. While Greeks accepted that
enormous footprints found in stone had been laid there by a race
of giants known as Titans, and medieval Europeans believed that
fossilized animal prints were the works of otherworldly creatures
ranging from devils to King Arthur's favorite dog, Cavall, the ease
with which fossils were found in North America made it difficult
to find supernatural explanations for something so commonplace.
(In what amounted to a last-ditch effort to find a biblical interpretation for what they were encountering in the New World, Puri-
tan colonists convinced themselves that the dinosaur tracks they
found were the footprints of enormous birds that had somehow
escaped from the menagerie of Noah's Ark.)

A French baron named Jean-Léopold-Nicholas-Frédéric Cuvier,
who was known as Georges, established the concept of extinction
in a 1796 lecture that refuted the commonly held notion that all
life fit into an unchanging Great Chain of Being that culminated
in God. After examining bones found in Siberia that included a
three-and-a-half-foot-long femur and several teeth that weighed
more than five pounds each, Cuvier argued that they corresponded
with no living animal, proving that life is made up of innumerable single flames that can be extinguished, rather than one solid
bonfire whose composition always remains the same. "What has
become of these two enormous animals of which one no longer
finds any living traces," Cuvier asked his audience, unveiling the
bones of what we now know as a mammoth. They "seem to me to
prove the existence of a world previous to ours, destroyed by some
kind of catastrophe."

Though he recognized that some forms of life have disappeared
from the Earth, Cuvier did not make the jump that those that
remained could evolve based on their environment, and argued
fiercely with anyone who did that anatomy cannot take half-steps
on the way toward a new species. In this, he made allegiances with
religious scholars who thought that the concepts of extinction and

evolution would inspire humankind to abandon all morality by
lessening the value of human life. "I am not quarreling or finding fault with a crocodile; a crocodile is a very respectable person
in his way," said the Reverend William Buckland, a theologian
known for his attempts to prove that fossils were the record of the
biblical flood and for accidentally eating the mummified heart of
Louis XIV of France because he thought it was a rare mineral.
"But I quarrel with finding a man, a *crocodile improved.*"

Owen enrolled at the Royal College of Surgeons in London in
1827, and was given the job of dissecting and identifying some of
the ten thousand animal specimens that had sat untouched in its
collection for over twenty-five years because someone had mis-
placed the manuscripts that labeled them. When Cuvier, whose
collection of prehistoric life at the National Museum of Natural
History in Paris had made him the most well-known of what were
then called fossilists, traveled to London to view the Royal College's collection of fossilized fish, Owen was named his assistant, in
large part because he was one of the few students who spoke fluent
French. They struck up a friendship, with the young, gangly Owen
becoming a constant companion of the obese, sixty-one-year-old
Cuvier, whom people unflatteringly called the Mammoth behind
his back. Owen was in his twenties and becoming a part of London's social and scientific elite, eventually raising his status so high
that Prince Albert would one day call on him to tutor the nine
royal children. Interacting with a giant in the field only increased
his ambition to rise higher at a time when the discovery of the
prehistoric world made it seem that humans were coming close to
fully understanding God's designs. Once, when he was introduced
as the "Cuvier of England," he complained in a letter that "I wish
they would be content to let me be the Owen of England."

The only problem was that others kept getting in his way. In
1811, the same year that Jane Austen published *Sense and Sensi-
bility*, a twelve-year-old girl named Mary Anning living about a

hundred miles away from her in southwest England unearthed the complete skeleton of an unknown bug-eyed creature below the cliffs at Lyme Regis that looked to have the head of a crocodile, the beak of a bird and the body of a giant, slender fish. Anning, whose father had died in debt and whose mother relied on charity to feed her children, sold the skeleton to a local lord for twenty-three pounds, a sum large enough to support the family for six months. The creature was soon taken to the British Museum, where English and French naturalists debated what exactly it was (the French soon lost patience with their British counterparts, with one of Cuvier's contemporaries declaring that English papers were "abstruse, incomprehensible and for the most part, uninteresting"). Charles Koning, who was then the Keeper of Natural History at the British Museum, named the animal *Ichthyosaurus*, meaning fish-lizard.

Curious what else Lyme Regis had to offer, naturalists began showing up at the Anning house and asking if Mary had anything new to sell them. She was often seen venturing below the cliffs wearing a long, dark dress, red scarf and white bonnet, clutching a pickaxe in one hand and a basket to hold her finds in the other. One friend who went on a prospecting trip with her later said that "we climbed down places, which I would have thought impossible to have descended had I been alone. The wind was high, the ground slippery, and the waves beating against Church Cliff. When we had clambered to the bottom our dangers were by no means over. . . . In one place she had to make haste to pass between the dashing of two waves. . . . She caught me with one arm round the waist and carried me some distance."

Anning's specimens were the chief attraction in a small shop her family opened a few blocks from the shore. It was there that she interacted with some of the wealthiest men of the era on a near-equal footing in the specialized realm of fossil discovery, a dizzying turnabout for a girl whom the local townsfolk had always found a bit odd even before she started pulling strange bones out of the ground—an opinion perhaps formed after she survived a lightning strike that killed three adults standing near her. "She frankly owns that the society of her own rank is become distasteful to her," one of her friends wrote.

Anning had neither the wealth nor the status to formally study the fossils she found, leaving others to claim the credit for her discoveries. She attempted to teach herself French in order to read Cuvier's essay about her specimens, desperately trying to connect with a world that shunned her despite her contributions. She was paid small sums for each specimen she unearthed, while well-connected scholars and naturalists used her work to brighten their own careers. Among them was Owen, whose attempt to flatter Anning by going on a prospecting trip with her was one of the very few times he set foot in a fossil dig in the field. Anning was guarded around him, and he left empty-handed. "She says the world has used her ill and she does not care for it," a friend wrote. "According to her account, these men of learning have sucked her brains, and made a great deal by publishing works, of which she furnished the contents, while she derived none of the advantages."

Anning was not the only one digging through England for fossils in hopes of securing a foothold in a better life. While she prospected in Lyme Regis, a shoemaker's son named Gideon Algernon Mantell could be found on the banks of the river Ouse sifting for what he called "medals of creation" that represented "the wreckage of former lives turned to stone." Mantell's preoccupation with the past stemmed in part from the fact that his now-lowly family could trace its lineage to a knight who had accompanied William the Conqueror in the Norman conquest of England in 1066. Sir Walter Mantell took part in a 1554 attempt to block the Catholic marriage of Queen Mary to Philip of Spain; when the plot failed, Sir Walter was among those executed by the monarch known as "Bloody Mary," and the family's estates in Kent,

Sussex and Northamptonshire were seized by the Crown. "In my boyish days I fancied I should restore its honors and that my children would have obtained the distinctions our knightly race once bore," Gideon Mantell wrote.

He became a country doctor, immersed in everything from treating smallpox and burns to amputating the legs of men and boys injured in the local mills, all while delivering between two and three hundred babies a year. When he was not healing others, he was a fixture at the Lewes library, devouring the work of James Parkinson, a doctor now best known for his description of Parkinson's disease but who at the time was famous for his work as a geologist. Mantell was drawn to Parkinson's attempts to square religion with science, such as his conclusion that the biblical story of Moses is "confirmed in every respect, except as to the age of the world, and the distance of time between the completion of different parts of creation." When Mantell came across a newly-opened quarry in the rolling, wooded hills of the Weald region of southeast England, which exposed layers of sediment up to forty feet deep, he climbed down into it and began collecting fragments of teeth and shells. He soon found the stump of what seemed to be a prehistoric palm tree and brought it home to add to what was becoming a small private museum.

He spent much of his free time chiseling through rock to unveil the fragments of what he soon identified as the femur and ribs of unknown animals. If not for the fact that he was holding the fossils in his own hands, he would not have believed that they existed, largely because of the size of the lifeforms they implied. One section of a rib stretched twenty-one inches long; a thigh bone measured almost thirty. The practicalities of sustaining such huge animals seemed impossible. How much food would something so giant need to consume each day? How could a muscle ever be strong enough to move the bones of a creature that was larger than a house? "I may be accused of

indulging in the marvelous, if I venture to state that upon com-
paring the larger bones of the Sussex lizard with those of the elephant, there seems reason to suppose that the former must have
more than equaled the latter in bulk and exceeded thirty feet in
length!" he wrote. "This species exceeded in magnitude every
animal of the lizard tribe hitherto discovered, either in a recent
or a fossilized state."

One morning in 1820 or 1821, Mantell brought along his wife,
Mary, on his medical rounds. While he was with a patient, she
passed the time by sorting through a pile of stones on the edge of
the road. Among the items she picked up was a smooth object the
color of mahogany that was more than an inch long and, upon
closer inspection, looked to be a fossilized tooth. When Mantell
returned, he immediately seized on the importance of her discovery. The tooth had a broad, flattened surface like those of mammalian herbivores, yet nothing in science at the time suggested
that mammals had lived in the prehistoric era. (Indeed, Parkinson's
essays argued that while the time elapsed in the Bible was off, the
order was not, implying that God had begun creating mammals
only after completing work on lesser forms of life, improving on
each one until he made humans in his own image.) The tooth was
unlike any fish, turtle or amphibian fossil, and seemed to imply
something else entirely. "As no known, existing reptiles are capable of masticating their food I could not venture to assign the tooth
in question to a lizard," Mantell wrote.

A tooth was not enough to prove the existence of an entirely
new form of life, nor were fossilized fragments of bone. Yet
together, they provided a wealth of evidence that was impossible
to disregard. Mantell, forever looking for a chance to restore his
family's past glory, grew convinced that he had discovered "one or
more gigantic animals of the Lizard Tribe" that were altogether
different from the type of fish-lizard uncovered by Anning. He
began sending the bones and teeth he found to eminent geologists

such as Cuvier, clawing his way into a social echelon from which he had long been excluded. "I am resolved to make every possible effort to obtain that rank in society to which I feel entitled both by my education and my profession," he wrote in his journal.

In his letters, he argued that he had found evidence of what he called an *Iguanodon*, a name derived from the fact that the tooth Mary Mantell found resembled those of an iguana yet was many times larger. When, in 1832, workers at a quarry in Tilgate Forest discovered fragments of petrified bone after blasting a particularly hard section of rock, Mantell bought the lot and had them carted to his home some thirty miles away. He spent weeks chiseling away until he could identify several vertebrae, ribs and a sternum, along with more than ten strange bones that were seventeen inches long and had no apparent purpose. Only after trying to place them in different sections of the body did he realize that they were in fact a form of armor that ran down the spine. He called the creature *Hylaeosaurus*, meaning forest lizard, and with that identified the first of what are now known as a family of armored dinosaurs called ankylosaurs.

With *Iguanodon* and *Hylaeosaurus*, Mantell was responsible for the discovery of two of the three species of what we now call dinosaurs that were known to science at the time (the other, a carnivore with serrated, bladelike teeth known as *Megalosaurus*, was found by Reverend Buckland at a slate quarry in Stonesfield, though he considered the find a minor accomplishment compared to his attempt to taste every living animal on Earth). Unable to shake the feeling that he might find a beast that measured over two hundred feet long, Mantell recentered his life around fossils, letting first his marriage and then his medical practice dissolve. As he rose up the social ladder, he received invitations to speak before the celebrated Geological Society in London, where he conversed with the sort of privileged men inclined to look down on country doctors who survived on the money they made delivering babies.

Among them was Richard Owen, who felt his chance at becoming the Owen of England slipping away. A proposal by the Geological Society to reclassify geological time based on the evidence of lifeforms, rather than type of stone, heightened Owen's desire to put his stamp on the biological record. In 1841, the layer of Earth's history that was once known as the primary rocks was rechristened the Azoic era, implying the absence of life. The layer of transition rocks was renamed the Paleozoic era, meaning ancient life, while the secondary rock layer became known as Mesozoic era, meaning middle life. The tertiary rock layer became known as the Cenozoic era, identifying it as the home of newer forms of life.

The pages of the history of life were waiting to be filled in, and Owen wanted to be the one who did it—or least received the credit. Though he never found any fossils of importance, Owen began to criticize Mantell's interpretations of the anatomy of *Iguanodon* and *Hylaeosaurus* and the similarities between species, and at times implied that he was the first to identify anatomical features that Mantell had previously observed. Mantell considered such acts "piracy" and resolved to no longer share his work with Owen.

In October 1841, Mantell was riding in a carriage across Clapham Common in London when the driver lost control of the horses. Mantell fell while trying to grab the tangled reins and was dragged along the ground, severely damaging his spine. Over the following days, Mantell felt a sensation of numbness spreading through his legs and soon found that he was unable to walk. His work scouring quarries and chiseling away stone appeared to be over.

Owen, meanwhile, continued to focus on the anatomical structures of Mantell's discoveries, and realized that the shaft of the *Iguanodon* femur was at a right angle to the pelvis, much like that of a mammal. By implication, this meant that it walked with its legs below it, like an elephant or deer, rather than with its legs splayed to the side, like a crocodile. Taking this line of thought further, that meant that the creature's tail—which, curiously, had proven difficult to find in the fossil record—would be drastically shorter than Mantell had estimated, leaving its overall size well below 200 feet.

In late 1841, a wealthy wine merchant and amateur geologist named William Saull purchased the first known sacrum, or lower spine, of an *Iguanodon* and installed it in his private museum. With Mantell unable to travel, Owen was one of the first to view it, and soon grasped its importance: the vertebrae of the lower spine were fused, just like the spine of *Megalosaurus*. That small adaptation allowed a creature's backbone to become incredibly strong, making it capable of supporting a body consisting of gigantic bones and mammoth muscles. The feature was not found on *Ichthyosaurus* nor on any of the other apparent sea creatures that Mary Anning had discovered, suggesting to Owen that there had once been a distinct group of enormous reptiles that, like contemporary mammals, were designed to walk upright on land with their legs tucked under them.

Though he excluded other prehistoric reptiles that would eventually become recognized as belonging to the same group, Owen focused on the similar anatomical characteristics of *Iguanodon*, *Hylaeosaurus*, and *Megalosaurus*, and argued they were large, unintelligent beasts that walked on all fours—a sort of reptilian rhinoceros. Over the next several weeks, he played around with different names for them, before settling on the term "dinosaur," which he introduced in an 1842 speech. With that, he claimed for himself the work that Mantell had done over the past twenty years, and within months attained the glory that would forever elude his rivals. The prime minister commissioned an oil portrait of Owen to hang in his home, and followed it up with a letter to the Queen recommending that Owen receive a royal pension equivalent to $20,000 a year in today's dollars. Mantel seethed in a letter to a friend that Owen had "altered names which I had imposed, and stated many inferences as if originating from himself when I had long since published the same," but had little recourse. His health continued to deteriorate, leaving him with few distractions beyond kindling his loathing of Owen. "It is astonishing with what intense hatred Owen is regarded by most of his contemporaries, with Mantell as arch-hater," said Thomas Henry Huxley, a prominent backer of Darwin at the time whose grandson, Aldous, would go on to write the novel *Brave New World*.

THE STUDY OF DINOSAURS REMAINED largely confined to England for the next thirty years until Arthur Lakes, an Oxford graduate who immigrated to the United States and later settled in Colorado to work as a schoolteacher, uncovered bones the size of tree trunks while on a walk with his friend Henry Beckwith in the spring of 1877. With a few minutes searching, they uncovered a vertebra nearly three feet in circumference that required the strength of three men to lift it into a cart. "It was so utterly beyond anything I had ever read or conceived possible that I could hardly believe my eyes," Lakes would later write.

He sent a few representative samples of the fossils to Othniel Charles Marsh, a professor at Yale whose fame rested on his perch as the first professor of paleontology in the United States. As a solitary child who had often clashed with his father after his mother's death from cholera, Marsh spent much of his time alone exploring minerals exposed by the digging of the nearby Erie Canal, and at the age of twenty received a portion of his mother's marriage dowry that allowed him to enter boarding school at Phillips Academy at the comparatively old age of twenty. He graduated from Yale University at the age of twenty-eight and led some of the first paleontology expeditions in the United States through what would become South Dakota, Nebraska, Utah and Wyoming. Hasty, possessive, and disinclined as he aged to go out into the field himself, Marsh held a chair at the university that was endowed by his uncle, George Peabody, who had built a fortune through setting up a transatlantic trade in commodities and a banking firm in London. Peabody's success made him one of the richest men in the United States, and his attempts to give most of his wealth away made him one of the most famous.

The gulf in personality between an uncle and nephew could not have been more vast. At his death, Peabody was praised for a life that was "illustrated and adorned by the constant practice of the most conspicuous probity, charity and good will toward mankind." Marsh, on the other hand, seemed inherently cold and suspicious, his temperament marked by "an absence of the complete exchange of confidence which normally exists between intimate friends," one of his longtime assistants would later write. "Even where perfect confidence existed, he seldom revealed more about any particular matter than seemed to him necessary." The seriousness with which he held himself was striking, even in an environment like Yale where self-importance was as common as a Gothic building on the three-hundred-year-old campus. "It was as if we had been two ministers of state having little acquaintance with each other, who had met for the settlement of some great question of public concern," wrote Timothy Dwight, a president of the university. "All was serious with a dignified solemnity, and measured with a diplomatic deliberateness." There was little daylight between his work and his conception of himself; he was a man with whom "close association . . . soon revealed both his unrestrained jealousy and his love of popular adulation," another contemporary noted. Had it not been for his uncle's support of Yale's Peabody Museum and his position as its director, Marsh would have been too disliked to rise higher, doomed to be forever a king in want of subjects.

The few friends he had did not stay friends for long. Among them was Edward Drinker Cope, a onetime professor of zoology at Haverford University whom Marsh met while studying in Berlin. The two remained in touch after they returned to the United

States, and, as if living parallel lives, both focused their attention on searching the continent for the sort of prehistoric bones unearthed in Europe. The similarities did not end there. Like Marsh, Cope had as a child found the study of the natural world a refuge, and later admitted that he was "not constructed for getting along comfortably with the general run of people." Despite his affinity for knowledge, he did not complete a university education. As a young boy, Cope could not meet the harsh standards of personal conduct at Westtown, an elite Quaker boarding school, and in his letters home often complained of demerits he had earned for talking too much or poor penmanship. He dropped out of school in frustration and shunned his father's offer to purchase him a farm, and instead pursued a scientific career without formal training. Throughout his life, he never lost the rough sheen which was at odds with his privileged upbringing; in time, one of his contemporaries would say that "Cope's mind was the most animal" he had ever encountered and "his tongue the filthiest."

The pairing of Marsh, who was forever on the alert for a slight that would remind him of the pain of his lowly childhood, and Cope, a man whose anger at not meeting the standards of his upbringing left him always ready to prove his worth, seemed destined to end in ruin. The first sign that their paths would clash came not long after Cope, who had quit his position at Haverford, alerted Marsh to fossils uncovered by diggers searching for the rich mudstone that was prized as fertilizer found in the marl pits of Southern New Jersey just outside of Philadelphia. At the time, just eighteen dinosaur species were known to have existed in North America, and most of those were identified only by an isolated tooth or vertebra—nothing like the nearly complete skeletons uncovered by Anning and Mantell. Cope invited Marsh to accompany him in a horse-drawn carriage on a tour of the pits near the town of Haddonfield in the spring of 1868. There, he introduced him to Alfred Voorhees, a local miner who often sent small bones he uncovered while digging to Cope in exchange for small sums. In the months after his tour with Marsh, however, Cope received no discoveries of consequence from Voorhees. When Cope inquired why the marl pits had apparently gone dry, he received what he felt to be evasive answers. Around the same time, Marsh routinely announced the discovery and acquisition of new fossils without revealing their source. Cope suspected that Marsh had gone behind his back and paid Voorhees to send the best bones to him in New Haven, though he had no proof.

The nascent mistrust between the two men turned darker the following year. Cope came into possession of more than a hundred bones discovered by an army surgeon in Kansas, and attempted to piece them together. The animal appeared to be the first known example of a marine reptile with an oddly flexible neck, the initial member of an order that Cope proposed calling *Streptosauria*, meaning twisted reptiles. Cope, who was twenty-nine at the time, prepared a paper along with lithographic plates to present to a major conference that summer and fully expected it to validate his work and make his name. Despite his suspicion that Marsh was undermining his ability to acquire more bones, he gave him an early look at his interpretation of the skeleton, seeking recognition from a man nine years his senior who was in a position to give him the respect he craved.

Instead, Marsh seemed to delight in pointing out all of the ways in which Cope erred, starting with the most embarrassing: Cope had put the head where the tail should be. There was no such thing as a twisted reptile. Cope instinctively rejected the critique before discovering for himself the mortifying fact that the skull fitted perfectly into the last vertebra of what he had thought was the rear of the animal. Deflated and angry, he tried to destroy all copies of the erroneous essay he had thought would make him famous. In its place, he found solace in the idea of humiliating Marsh as retribution. "His wounded vanity received a shock from which it has

never recovered, and he has since been my bitter enemy," Marsh
later wrote.

For the next twenty years, each man attempted to find more
fossils, name more species and write more papers than the other,
a feud that would later become known as the Bone Wars. Arthur
Lakes's discovery of enormous fossil specimens opened a new
front, taking the battlefield away from the marshes of the East
Coast and into the canyons and cliffs carving the western half of
the country, a land that was newly accessible through the construction of the Transcontinental Railroad. There, everything
seemed greater: fossils which suggested dinosaurs so big that they
stretched imagination; the starkness of the landscape; and the prize
of finding what very well could be the largest animal that ever
walked on Earth.

In an untamed land, Marsh and Cope raced to find as many
bones as they could, a pursuit that was equaled only by their
attempts to prevent the other from succeeding. They paid off
informants; spread lies about the other; planned to dynamite
quarries rather than have them fall into the other's hands; publicly wished the other was dead. Both fanatically tracked news
of any fossil discoveries published in small newspapers in the
West, relying on the newly-built railroads to crisscross the country in hopes of a new lead. Marsh, with his abundant wealth
and resources, often had the upper hand, directing teams of Yale
students to open numerous quarries each summer and move on
quickly if they did not immediately find something worthwhile,
a policy built on the idea of strength in numbers. Seeking an
advantage where he could find it, he made an agreement with
Red Cloud, an influential chief of the Oglala Lakota tribe, to
lobby Congress in defense of Native American sovereignty in
exchange for the exclusive right to prospect in an area of what is
now Montana and Wyoming. Even with the field titled his way,
Marsh was not above playing dirty. On one of the rare occasions

when he found himself prospecting near Cope, Marsh snuck into his rival's digging site at night and scattered unrelated fossils in hopes of confusing him.

The pettiness touched nearly everyone connected with both Marsh and Cope, often making it hard to tell whether discovering fossils or denying the other glory was each man's true goal. Their assistants took on the conflict as their own, like sin passed on through generations. At a rural outpost in Wyoming known for its abundant fossil beds, prospectors working for Marsh sent spies into a camp of men working for Cope. In return, Cope's team locked their rivals out of a train station in order to prevent them from sending their haul back to Marsh on the East Coast. Before long, the opposing camps were throwing rocks at each other, as if all of Wyoming were nothing more than a sandbox.

Cope, consumed by his need to best Marsh, compensated for his dwindling financial resources with a fervent drive to publish, writing more than 1,400 scientific papers on topics ranging from dinosaurs to early mammals. Soon, his collection of fossils was the only thing of worth he owned, his inheritance destroyed by a consuming desire to beat the one person who fully understood his need to find and discover new species. His obsession with proving his superiority prompted Cope in his will to donate his brain to science and challenge Marsh to do the same, convinced that whoever's brain weighed more would be proven the true intellectual superior.

News of their discoveries appeared regularly in the largest newspapers on the East Coast, seemingly confirming every suspicion that the West was an alien landscape where the bones of enormous monsters could be found sticking out of the ground. Between them, Marsh and Cope discovered and named more than 120 new species, including some of the most familiar dinosaurs we know now, such as the bony-plated *Stegosaurus*, the long-necked *Apatosaurus*, a name meaning deceptive lizard for the way some of its bones resembled an unrelated aquatic reptile, and its closely-related cousin *Brontosaurus*, meaning thunder lizard.

In the process, they created a commodity: because of the well-publicized rivalry, farmers realized that scientists would pay out-sized sums for the fossils they often uncovered while clearing land, which provided an incentive to keep and protect what were previously considered novelties at best and nuisances at worst. The new market was not lost on men throughout the West who had failed to find gold or silver and were brave enough to explore unforgiving regions. A new profession—fossil hunter—materialized out of the Bone Wars, like an unexpected side effect of mixing caustic chemicals. On a scale never imagined by Mary Anning, relics of prehistoric life were taken out of the domain of academic essays written by a small community of naturalists and reimagined as precious materials that would make anyone who found them rich.

To narrow the gap between his own published output and Cope's, Marsh began putting his name on papers written by his assistants, ruining what little loyalty they felt toward him. Williston, who would one day become Barnum Brown's professor at Kansas, was among those who found their careers blocked by Marsh's habit of taking credit for others' work. He wrote a private later to Cope complaining that Marsh "has never been known to tell the truth when a falsehood would serve the purpose as well." Cope shared the letter with a newspaper reporter in an attempt to discredit Marsh, severing Williston's ties with Yale.

Williston returned to Kansas as a professor in 1890 and prepared himself for his first expeditions free of the shadow of Marsh, whose own status had been damaged by the scandal of his behavior toward Cope. Funding from the U.S. government to mount further expeditions suddenly became scarce, and Marsh retreated further into the sanctuary of the Peabody Museum, where he remained unrepentant and fearful that someone would one day find dinosaurs that made his achievements look small by comparison.

For the first time in his life, Williston had the money and time to launch his own expedition, blessed in part with a location in Kansas that kept costs lower due to its proximity to great fossil beds filled with the unknown. Through their feuding, Cope and Marsh had effectively sabotaged themselves, making others unwilling to work with them and clearing the stage for the start of another era, like a new layer of sediment laid down on top of their discoveries. Williston had seen some of the great fossils lying scattered in the West, and knew that others must be out there just waiting to be found. All he needed were some students to supply muscle.

Chapter Three

SCRAPING THE SURFACE

THROUGHOUT HIS LIFE, BARNUM BROWN COULD BE FELT as much as he was seen. Immensely strong and buoyed by deep reservoirs of energy supporting his lean frame, there seemed to be no task that could discourage him, as if he were forever a boy stuck in the body of a man. That was not the only thing that was childlike about him. He seemed to pay no attention to hierarchy or rank, looking past the trappings of social status without ever feeling the need to question whether he belonged. If he was interested in something, he would go for it, not stopping to think of all the reasons why he shouldn't. When he heard that Williston was planning an expedition to the White River Badlands in South Dakota in the summer of 1894, he talked his way onto the nine-person crew despite never having taken a paleontology class. He was twenty years old and wanted nothing more than to continue what felt would be a life of ever-increasing fun, free to pursue where his interests pulled him.

The party left Lawrence on June 13, 1894, reaching the badlands nine days later after taking a boat up the Missouri River. In an echo of his trip through the disappearing West with his father, Brown was called on to do far more than anyone else. He cooked the expedition's food, washed their clothes and did all the other small jobs that were required to keep a pack of ten men alive as they searched for fossils under blistering skies that at any moment could be shattered by torrential thunderstorms and hail. In return, he learned how to turn his raw energy into skill.

Paleontology as a professional endeavor was barely fifty years old, and still to a startling degree relied on luck. The fossils that inspired Owen to invent the term "dinosaur" had been found by accident, unintentional byproducts of mining operations and road construction, and were given scientific value only when they fell into the hands of someone who recognized what they were. Anning, meanwhile, concentrated nearly all of her prospecting within the small region where she happened to grow up, never having the money nor the opportunity to venture to new locales in hopes they would prove as bountiful as home. Marsh and Cope may have known how to identify bones when they were brought out of the ground, but had very little skill at discovering new fossil beds themselves. The betrayal of paying off the other's sources of bones was what truly fueled the rancor of their twenty year-long battle, a reflection that both men remained hopelessly dependent on tips from those who dug into the earth for some other purpose and came up with the unexpected, like reeling in a shoe when they were trying to catch trout. Aside from miners and farmers, most of the fossils found during the Bone Wars were discovered by hikers, who came upon gigantic bones sticking out of the ground by happenstance, and cowboys, who noticed what looked like strange bison skulls sticking out of the ground while driving cattle through canyons. (One early prospector sold several items to Marsh after he realized that ants on the prairies pick up and carry small, hard objects to form mounds around their nests, leaving anthills ringed by the jaws and teeth of dinosaurs.)

Venturing into the field with the expectation of finding fossils that no one knew were there was in many ways still a novelty. Fieldwork, when it was done, consisted of taking the knowledge gained from hard labor and applying it to science. Paleontology required drawing from both physical strength and intellectual skill, and those who could bridge the divide inherent in the profession would soar to its greatest heights. On those dusty, hot days in the badlands, Brown received his first lessons in how to dig with purpose, learning to manipulate tools that were the dominion of miners and others skilled in breaking apart the earth without destroying the treasures hidden within. It was a welcome respite from the classroom for a student just then realizing his academic limitations. Here, knowledge came through exertion, the product of hours stooped over in the sun muscling closer to a potential specimen only to find that there was still more work to do, like opening a series of Russian nesting dolls.

Pickaxes and shovels were suitable for taking apart big rocks; rock hammers were the most useful when working in the neighborhood of a bone; digging knives and trowels were required when you were close enough to extract it. Anything more delicate than that required a pocket knife. Paintbrushes kept an area clean, and plaster would stop a newly excavated bone from crumbling once exposed to air. Dynamite was handy to clear an area, but use it too liberally and you'd blow up what had sat untouched for millions of years before you came along. If blasting was impossible, then a scraper plow attached to a team of horses could take off a layer of twenty feet of rock and dirt. And there was nothing wrong with keeping the fossil encased in rock once you found it; that's what the hammers and chisels back in the campus laboratory were for.

Once Brown learned the fundamentals of how to dig, he needed to learn the art of where to direct his newfound ability. Honing the talent to tell fossil from stone is something close to acquiring a new sense; in these often harsh environments, the blanching of the sun can make it impossible to rely on color or weight alone. A conical shape may hint that the object in question is a tooth, while a curve or an unusually straight line may be the first sign of a bone.

If it is still a mystery whether you are holding a rock or a fossil in your hand, there is one last resort: licking it. The tip of the tongue briefly sticks to fossils, yet will not stick to stone. (Licking is also a go-to for geologists in the field hoping to make an easy identification of what minerals they are working with: quartz, calcite and gypsum often look remarkably like halite, the mineral commonly known as rock salt, but do not have the same distinctive flavor; chrysocolla and kaolinite, meanwhile, are known to be remarkably sticky and also tasteless.)

The confirmation that an object is indeed a fossil simply leads to more questions. Depending on which part was exposed, it can often be a matter of guessing which way the skeleton continues on beneath the shell of rock and dirt, the combination of which is known in the field as its matrix. Sometimes, what seems like the logical path of a bone ends in nothing, requiring you to retrace your path to discover the point where it jutted off, if it did at all. Other times the bones seem to plummet deeper into the earth, forcing a calculated guess as to whether launching a further assault is worth it. That is not all. What looks like the promising start of a femur or mandible will sometimes turn out to be nothing but the indistinguishable nib of a toe—a disappointment compounded by the fact that it took four days of punishing labor to confirm the fact that you are no better off than when you started. As a whole, the work of finding fossils amounts to moments of joy surrounded by days of toil, an unbalanced equation that often pushes away those who are unwilling to shoulder the near-certainty of repeated failure.

As soon as he got into the field, Brown seemed to have an innate sense of how an animal's body was positioned at the time it was buried and fossilized, a feel for the juxtaposition of rocks and the relics of life that no amount of anatomy courses could teach. With a glance he was often able to tell whether what appeared to be a knob of rock was a glimmer of a larger knee bone, or whether a seam would curve and reveal itself to be a spine. He had already distinguished himself through his physical brawn; now, it was his ability to take on what another team member had started and quickly determine whether it was worth the effort that made him invaluable. It was if he had a magical ability to unearth a specimen, like someone who can sit down and complete a jigsaw puzzle without first needing to find the edges. Sometimes he would disappear into picked-over quarries and come out with fossils everyone else had missed, a feat that happened with enough frequency that it seemed like he could see layers beneath the surface.

The harder question was where to start digging in the first place. Finding a fossil requires the ability to see, outside the bounds of time, two different scenes at once: the immediate features of the landscape before you, and what it likely looked like millions of years ago. Geological maps which identified the type and age of rock were relatively rare in Brown's era, meaning that many expeditions were simply a search for the Mesozoic-Era sedimentary rocks that had the potential to hold dinosaur fossils. Actually finding a specimen on some of these scouting missions was considered an unexpected stroke of fortune.

Searching for fossils took expedition parties into some of the most difficult terrain on Earth, picking their way over rock formations that are the leftovers of once-verdant landscapes. The snaking red-striped sandstone ravines of the badlands, their ridges jutting out like the Earth's bones, were once lapped by ancient streams and lakes. Animals that died either in or near the water could become covered by mud and silt, starting the process of fossilization. Over millions of years, the weight and pressure of additional layers of sediment turned the remains of once-living creatures into stone, and in some cases lifted rock formations that were once at the bottom of vast inland seas or rivers into canyons and gullies. A paleontologist on the hunt for a fossil first looks for a subtle variation of colors, which can be a sign that different types of sediment came together in an event such as a flood that could have buried the remains of an animal within it. After color, the next sign of a fossil is often texture. Some bones are marked by countless little holes, as if they were aerated, while others are smoother and shinier than rock alone. Adjusting to the barren landscape of the fossil beds, with its narrow band of colors, required becoming comfortable in a land that seemed so forbidding as to never have harbored life.

Not everything about being in the field was work. The badlands into which Williston led his crew that summer were filled with crews from other colleges, making the experience something closer to a raucous spring break than the ruthless competition between Cope and Marsh. "Princeton girls did not come today which was a great disappointment after greasing our shoes and washing up," Brown complained in his diary one night, most likely referring to a crew from Evelyn College, the university's short-lived women's college. (Princeton itself did not admit women until 1969.) "Saw a beautiful sunset. In the evening drowned out [a] Nebraska [crew] with 'Carmine' and other familiar college songs." Later that summer, Brown's team "ushered in the glorious fourth of July with twenty or thirty shots which made Neb think the Sioux were upon them."

Brown's first find was the skull of an *Oreodont*, a piglike mammal that lived in abundant numbers on the Great Plains five million years ago. "The Dr. is well pleased with my skull. It is the greatest find so far," he boasted in his field journal. Williston soon began to rely on Brown not only for his strength—few others in the team could manage the job of moving delicate, heavy objects up and down the ravines without dropping them—but for his patience. Fully unearthing a fossil required digging around the specimen and covering it all with hardened flour paste, which formed a protective jacket until all of the remaining rock could be carefully chiseled away in a laboratory. Williston did not trust himself to wait. "He would start to work on a specimen and then turn it over to me and say, 'Brown, you take it out. I am so anxious to see the specimen when it is out that I am afraid I will injure it in excavation,'" Brown wrote. Over the course of a month and a half, the team found the remnants of a saber-toothed tiger and other prehistoric mammals, but failed to find the bones of a dinosaur.

The following summer, Williston again tapped Brown to work as a field hand on a trip to Wyoming in search of the skull of a *Triceratops*. The first known specimen of the species—whose three horns and parrotlike beak gave it one of the largest heads of any known land animal—had been found in the region six years earlier when a cowboy named Edmund Wilson saw what he thought was the head of a steer hiding behind some rocks on the bank of a gulch during a roundup and threw a lasso over it. Ranch hands who rode alongside him described it as having "horns as long as a hoe handle and eye holes as big as your hat." Upon closer inspection, Wilson realized it was the skull of a creature that looked like an alien. When he pulled the rope to try to drag the skull with him back to the ranch, it rolled to the bottom of the gulch, leaving just one of its horns intact. Word of the find soon reached Marsh in New Haven, and he sent an assistant curator by the name of John Bell Hatcher out to Wyoming with instructions to bring back the skull as soon as possible. When it reached Yale, Marsh determined that the creature's head weighed half a ton. He quickly wrote the first formal description of the animal, which appeared in the *American Journal of Science* in April 1889. Marsh did not know what to make of the creature, calling it a "strange reptile" and noting that "other remains received more recently indicate forms much larger and more grotesque in appearance," before bestowing upon it the name *Ceratops horridus*.

Other *Triceratops* specimens—which were also originally known by the competing name *Ceratops*—popped up over the next several years in the same region, leaving one prospector to call it the "*Ceratops* beds of Converse County." (The fossil beds, which spread through eastern Wyoming and North Dakota, are now known as the Lance Formation. With some sections nearly two thousand feet thick, they have accounted for some of the most spectacular finds in the history of paleontology.) As the specimens were relatively plentiful, the problem was not finding them; rather, it was getting one out of the field without destroying equipment in the process. The skull alone of one specimen discovered in North Dakota weighed more than three tons, requiring a team of horses to haul it out of a ravine and several broken-down carts to pull it across the prairie.

No one had ever put a *Triceratops* on display, however, and Williston hoped that by finding one he could bolster the small museum at the university while also giving himself the added pleasure of annoying Marsh, who had lost the federal funding that he relied on to fund his expeditions in 1890 due in large part to the fallout of the Bone Wars scandal. "I shall await the results with interest," Williston wrote to a colleague after learning that Marsh was working on a book based on the fossils Williston had helped collect. "Perhaps, if I get the material from a brand new locality that I am after, [I can] undo some of his work."

Brown helped lead a team of horses north from Kansas, arriving near Lusk in the first week of July 1895, just in time for a raucous Fourth of July. "From all directions the cowboys are coming to begin the celebration," a worried university trustee who accompanied the prospecting team wrote in his journal. "Very soon a large bonfire is started in the middle of main street and the sound of all kinds of fireworks begins. The fellows running their horses through town, full tilt, fire their revolvers into the fire just to see it fly. They keep up this orgy nearly all night and we get but little sleep."

Brown was raised in a town where saloons were the only source of entertainment, and was the rare person who felt comfortable in both the rough world of the frontier and among scientists who only ventured into the badlands to excavate fossils they hoped to place in a museum. The longer he stayed in the field, the more he realized that this—the open lands, the adventure and the mixture of science and the frontier—was his natural element, a place where he felt fully free.

He soon proved himself worthy. Within two weeks of arriving in Wyoming, he helped find two *Triceratops* skulls—each one six feet long, four feet across, and three feet thick—embedded in a high sandstone bluff deep in the badlands. Each specimen was among the finest ever recorded, their refinement a testament to Brown's ability to dig but not destroy. By the time the expedition returned to Kansas, they had collected an additional five tons of fossils, and Brown had built a strong enough relationship with Williston that the professor invited him to live in his house during the fall semester. "[Brown] took a great deal of the care & looked after things in camp. Things others wouldn't think of," Williston's wife later remarked in a letter, singling him out from a cadre of students and professionals who shared the same dreams.

With the expedition a success, Williston began preparing for the following summer's prospecting season. He received a letter from a colleague at the American Museum of Natural History in New York, asking if a student whom he had taken with him to South Dakota might be available to work the following summer as an assistant. Williston, however, suggested a different person for the job.

"Brown has been with me on two expeditions, and is the best man in the field that I ever had. He is energetic, has great powers of endurance, walking thirty miles a day without fatigue, is very methodical in all his habits, and thoroughly honest," he wrote, before continuing, "The man whom you remember in Dakota was probably Dickinson. He was a very good student in the University and a good fellow, but a complete failure in the field."

Not long after, a letter bearing a Manhattan postmark arrived for Brown. Inside it sat an invitation that changed the course of his life.

CREATURES
EQUALLY COLOSSAL
AND EQUALLY STRANGE

THE LETTER BROWN HELD IN HIS HAND WAS A PORTAL TO a better life, an invitation that could not have seemed more out of place than if Fifth Avenue itself had somehow crashed onto the dirty Kansas plain. In any other context, for any other reason, the lives of a scion of Manhattan society and a man who had spent his childhood playing on mounds of coal would have remained forever distant, like opposite poles of the Earth. And yet Henry Fairfield Osborn was not used to being denied.

The eldest son of a founder of the Illinois Central Railroad, then one of the nation's most profitable and powerful companies, Osborn grew up in the gilded cocoon of New York City's aristocracy in the midst of the country's greatest economic boom. When he was a year old, his uncle John Pierpont Morgan, the most important man on Wall Street, would stoop down and play with him on the family's parlor rug. By the age of three, the boy would walk into a room and start reciting facts or sermons and expect his audience to stop whatever they were doing and pay attention. "Every day he talks up something new," his mother noted. During the summer, he explored the family estate on a mountaintop in the Hudson Valley directly across from West Point, often swimming in the river with his younger brother's childhood friend Theodore Roosevelt. The rest of the year he spent at the family's four-story brownstone mansion at 32 Park Avenue in Manhattan, where boxes of pansies and tulips hung from each windowsill in the spring and landscape portraits by Frederic Edwin Church, perhaps the most famous painter then living in the United States, decorated the walls.

When his father was increasingly called away on business after the Civil War, Osborn assumed command. He began acting as the head of the family as a teenager, controlling everything from the household expenses to travel arrangements for his siblings. Whatever shyness was still lurking within him was expelled at the age of fourteen, when his father demanded that he begin publishing a newspaper that he called *The Boy's Journal*, which forced him to interact with adults as equals. Later that summer, the family toured Europe, relying on Osborn to manage all the concierges and train conductors they came in contact with, his innate sense of superiority their solution to any puzzle. When it was time to choose a college, he diverted from the pipeline of boys at his prep school who ended up at Yale and went instead to Princeton. His maternal great-uncle, the Reverend Ebenezer Pemberton, had been one of its three founders before the American Revolution.

At Princeton, Osborn experienced the first conflict between the young prince he thought he was and the person others perceived him to be. He arrived at the all-male campus with slender limbs, short-cropped hair and a high collar, a look that one classmate described as "almost girlish" and which soon translated into a nickname he detested, "Polly Osborn." Attending all of the school's football games and joining the sculling team did little to improve his social standing. Only after he won the annual cane spree— a campus tradition in which freshmen and sophomores brutally attacked one another on a playing field with bamboo canes until only one bloodied man was left standing—did he begin to earn the respect of his classmates. Any doubts that entered his mind as to who he would become were drowned under constant letters from his mother, Virginia, that reminded him of his assured place in life. "Make up your mind to lead in college . . . not be led by the more worthless half of class," she wrote. A deeply religious woman who often cited Bible verses in conversation, she expected her children to suffer to become closer to God, and was happy when she learned that because of a problem in the dorm building her son's room had no heat in the winter. Over time, Osborn began to believe that he was destined for great things, having never been exposed to anything that would give him doubts. The fact that he was not an especially brilliant student and eventually graduated in the middle of his class did nothing to dent his conception of himself.

Not that his grades would matter in any real sense. His path in life had already been cleared by his father, who expected that his son would join him in the railroad business as soon as he made the social connections among his classmates that were in many ways the whole point of being there. Free to explore without worrying about his future job prospects, Osborn in his junior year took his first course in geology, which was taught by Arnold Guyot, a professor who had founded the U.S. Weather Bureau as well as Princeton's own small natural history museum, and whose influence on his field was so great that guyots—naturally occurring flat-topped mountain peaks that rise from the ocean floor—would eventually be named after him.

It was in those classroom sessions that Osborn was first exposed to the revolution happening right under his feet. He read reports about the fossils collected by Marsh and his assistants at Yale and became fascinated by the remnants of the Earth's history, seeing in the slow engines of geology and evolution a confirmation of his conception of God. Until the end of his career, Osborn would never break from the Presbyterian faith that his mother honed in him, arguing in one college essay that eternal salvation could only be attained through unceasing effort. Yet he also clung to an older, nearly Calvinist view of predestination that animated his sense that he was of a higher caste than nearly everyone he came into contact with. "There is no indication of a predisposition on my part to this life of research; in my school and early life I was never conscious of such a predisposition nor am I able in a review of my ancestry and of my own boyhood to account for my life vocation," Osborn wrote in an essay soon before he retired. "My boyhood and youth were similar to those of almost any other boy, undistinguished by a display of the driving force which from the moment of its awakening in the junior year of my Princeton days, has ever impelled me with constantly increasing power. The impulse which led me to dedicate my life to research must truly have come from within."

Marsh was one of the first to try to stand in the way of Osborn's desires, flicking off his presumptuous request as a college student to view some of the professor's fossils that had not yet been announced through the publication of a scientific paper. When Osborn arrived at Yale to see its collection in person, Marsh secretly followed him and a group of fellow Princeton students around the Peabody Museum in his slippers to ensure that they stayed in a small, pre-approved area. Not to be outdone, Osborn helped organize what would become known as Princeton's first Geological Expedition, in which eighteen students and two professors embarked on an eleven-week trip through Colorado, Wyoming and Utah in the summer of 1877. They hired two private train cars—one to travel in, the other to hold their luggage—and headed for the open West, stopping first to sample the metropolitan offerings of Chicago and Kansas City.

The expedition went about as well as expected, given that it consisted of a group of young men whose idea of roughing it consisted of arguing with waiters in Europe. "Of our journey, novel to most of us though it was, there was not much to be said," wrote William Berryman Scott, a fellow junior who was the son of a professor at Princeton's theological seminary. "The Middle West was not then the busy, prosperous region it has since become, and the principal impression which it made upon me then was one of crudeness and shabbiness. The roads were quagmires of black mud; the towns were chiefly of wood and sadly in need of paint and, though there were a great many fine-looking farms, the journey was a depressing experience."

The group did not find any fossils of note, but returned with their enthusiasm for the natural world undaunted. Osborn recast his life around science, ignoring his father's pleas to consider a more lucrative line of work. He spent two years of postgraduate work in New York and Princeton, and, thanks largely to his connections, was named an assistant professor of natural history at Princeton in 1881. Like most of the things in his life to that point, Osborn's success would not have been possible without the financial backing of his father, who first donated money to expand the college library while his son was completing his graduate work and then, when the failure of the university's hotel threatened the funding of several professorships, agreed to pay Henry's full salary, floating his son's ambitions on a cloud of family money.

Despite his advantages, Osborn did little to distinguish himself as a scientist. He initially dove into embryology, determined to prove how the marsupial yolk sac conveyed nutrients to a fetus. After several years of failure in the best laboratory on campus, built with donations from his father, he turned his interest toward psychology. When that proved futile, he made plans to found and edit a scientific journal that never got off the ground. Finally, after four years on the faculty, he pivoted to vertebrate paleontology and proposed coauthoring a textbook on North American fossils with Scott, his onetime companion on the Geological Expedition, who had already built a reputation in the field while at the same time shouldering the load of teaching all the paleontology courses at the university. Osborn knew little about the subject but, thanks to his ability to control the pipeline of funding, he expanded the paleontology department and cast himself in the role of supervisor. The position allowed him to take credit for his colleague's work, a pattern he would replicate over the remainder of his career. He settled into what he considered his rightful place and began crafting theories on how the immense fossils then turning up in the American West fit into the concept of evolution and the emergence of mammals and humankind, relying almost entirely on collectors and other field scientists to do the hard work of finding bones that he would then examine in the comfortable confines of his laboratory.

Still, he longed for bigger things. His father constantly reminded him that Princeton was neither a major social nor scientific center of the country and that every day he spent raising his four young children in New Jersey was a wasted opportunity, given their proximity to the connections that could be forged in Manhattan. Osborn, too, craved the prestige that could only be satisfied with a life centered in New York City, and often traveled with his wife and children to socialize with the city's elite at his parents' summer mansion, named Castle Rock, which overlooked Garrison on the Hudson. In the late 1880s, he began making his desire to move back to New York known to members of his father's social circle, and in 1890 he was offered a position at what is now known as Columbia University by Seth Low. The heir of a prosperous trading firm known as A. A. Low and Co., Low was perhaps more famous for serving a term as the mayor of Brooklyn, where he pushed for reform in the city's public schools. He knew that by adding Osborn to the faculty the university would stand to benefit from the wealth not only of his family but of their friends as well, and Osborn soon headed a new department of biology.

Osborn, however, had other things on his mind than trading one university position for another. While securing his professorship at Columbia, he convinced Morris Jessup, a former railroad titan and friend of his father who was now the president of the fledgling American Museum of Natural History, to bring him in to lead the museum's new Department of Vertebrate Paleontology. Though it was a small department at an institution that barely registered in the minds of New Yorkers at the time, Osborn saw in the position a way to bolster his own reputation by transforming the American Museum into a player in the rapidly expanding world of paleontology—perhaps the one branch of science in which Americans were held in higher esteem than their counterparts in Europe. For that, he had to thank the man who had stood in his way. "There is nothing in any way comparable . . . for their scientific importance, to the series of fossils which Professor Marsh has brought together," noted the British evolutionist Thomas Henry Huxley after touring Yale's Peabody Museum in person.

By positioning himself as the heir to Marsh and Cope without the weight of scandal, Osborn envisioned a path toward becoming the most prominent scientist in the country, able to exert his intellectual influence in the same way that his father's millions remade his material world. The board of trustees soon voted to provide Osborn what he needed to make the paleontology division further "the cause of science, education, and popular interest" at the museum. To do so, Osborn would need to build up a division from scratch. The museum was a mass of empty rooms, and it was his job to fill them. For the first time in his life, Osborn would have an independent measure of his success or failure that could not be swayed by his father. In order to succeed, he needed to revolutionize a museum in which the public had shown little interest. And for that, he needed a person that he could trust to bring him the fossils and subsequent fame which he considered his due.

✦ ✦ ✦ ✦

DESPITE THE IMPORTANCE OF DINOSAUR fossils to the growing understanding of the history of life on Earth, very few people had actually seen one. The conception of what a dinosaur looked like when its bones were set back into their rightful places was at best provided by drawings done by naturalists that were rooted in science but filled in with fantasy, and at worst a fancy of the mind no different than daydreams of a dragon. How the muscles and bones of a living dinosaur shaped its body required a leap of imagination, like conjuring the finished form of a ship after discovering its scattered wreckage at the bottom of the sea.

Nevertheless, in the early 1850s, Prince Albert, in a nod to Britain's outsized role in their discovery, decided it was time to build the world's first life-size models of dinosaurs. He turned to Sir Richard Owen, who had come up with the classification of dinosaurs less than a decade before, and work soon began on an exhibit depicting the prehistoric world that would serve as the centerpiece for the Crystal Palace, an amusement park, zoo and ornamental gardens planned for Sydenham, a suburban section of southeast London. Owen envisaged a collection consisting of models of the three known dinosaurs at the time situated around a lake, as well as models of flying reptiles known as pterosaurs and giant mammals such as a sloth and a deer, whose antlers would consist of actual fossils. To make the idea a reality, Owen hired a sculptor named Benjamin Waterhouse Hawkins, a former assistant superintendent of the London World's Fair best known for making models of living animals for the Earl of Derby.

With nothing other than Owen's drawings and theories to go on, Hawkins cleared his workshop and began constructing the first life-size forms of dinosaurs to appear on Earth in millions of years. He began by crafting small models out of clay, basing the way that each animal's limbs and muscles hung from its body on Owen's conception that dinosaurs were large, lumbering, dimwitted beasts that mainly lived in or near water in order to provide relief from the strain of moving their enormous bodies. Once each model— the most ambitious and accurate depiction of dinosaurs ever attempted at that point—met Owen's approval, Hawkins turned to brick, iron and cement to forge full sculptures that when finished weighed up to thirty tons, a process he called "not less than building a house upon four columns." Sea lizards, *Iguanodons, Megalosaurus* and pterodactyls began to take shape in his South London workshop, slowly bridging the abyss of time between the Paleozoic Era and the Victorian Age. Each scaly creation was painted a shade of light green, making it appear to be the unintended result of mating a garden lizard with a particularly tall and fat hippopotamus.

To build up the public's interest in the new park, Hawkins invited twenty-one of London's leading scientists and newspaper editors to dine on an eight-course feast served inside the *Iguanodon* mold on New Year's Eve of 1853. Owen sat at the literal head of the table, greeting each person as he climbed up a small ladder to reach the belly of the beast. The party went long into the night, with speeches heralding Owen's achievements, and continued as the distinguished group of men made their way to the railway station. Their drunken chorus was "so fierce and enthusiastic as almost to lead to the belief that the herd of *Iguanodons* were bellowing," Hawkins wrote.

The coverage of Hawkins's publicity stunt was the first widespread intrusion of dinosaurs into the public consciousness, sparking one of the first social manias based in science. Hawkins was besieged with requests to tour his workshop before the official opening of the Crystal Palace, and started a lucrative business selling miniature casts of his models. Charles Dickens wrote letters to Owen begging him to write for his journal *Household Words*. When Queen Victoria opened the Crystal Palace on June 10, 1854, forty thousand people lined up outside. Sir Richard Owen stood with the French emperor and the king of Portugal as the Queen, clad in a light blue dress and white shawl, gave a speech in which she said she hoped "that this wonderful structure, and the treasures of art and knowledge which it contains, may long continue to elevate and instruct, as well as to delight and amuse, the minds of all classes of my people." Hundreds of thousands of visitors toured the grounds of the Crystal Palace over the next decade. Those that did not make the trip themselves could hardly miss the posters and models of Hawkins's creations that were widely sold throughout the country, spurring the imaginations of writers including Jules Verne and Louis Figuier, whose work soon featured dinosaurs sparring with one another.

The models made the existence of dinosaurs real in a way that drawings of fossils could not, upending the conception of the world that was then being handed down by education and religion. It was as if several pages had suddenly been added in the middle of a familiar story, forcing readers to reconcile everything they now knew with what they had always believed. Though science had by this time disproved the biblical story of a great flood, belief in it remained so widespread that a London weekly magazine ran an essay devoted to Hawkins's models which postulated in all sincerity that these "savages and beasts" had become extinct "because they were too large to go into the Ark, and so they were all drowned," without asking why Noah was not instructed to build something bigger from the get-go.

The worldwide popularity of Hawkins's models in Crystal Park led the commissioners of the new Central Park then under construction in New York City to want an expanded version of their own. As it emerged from the Civil War, New York found itself becoming the world's busiest port and one of its wealthiest cities. Yet it continued to feel inferior to its rivals, London and Paris, and an exhibit featuring dinosaurs discovered in the American West—larger, heavier and presumably fiercer than anything found in Europe—would be a fitting way to underscore the superiority of the New World. While he was on a speaking tour of the United States, Hawkins received a commission to create models for a planned Paleozoic Museum that would stand within the park, attracting millions and enhancing New York's reputation with its own wonders to display. "For thousands of years men have dwelt upon the earth without even suspecting that it was a mighty tomb of animated races that once flourished upon it as the living tribes do now," the park's commissioners wrote in an annual report announcing their plan to build a dinosaur-focused museum. "Only in very recent times, which men still remember, was the discovery made that earth has had a vast antiquity; that it has teemed with life for countless ages, and that generations of the most gigantic and extraordinary creatures lived through long geological periods, and were succeeded by other kinds of creatures equally colossal and equally strange."

In the ten years since Hawkins built the Crystal Palace sculptures, scientists—many of them animated by the opportunity to correct a man who was widely despised—had realized that there were fundamental flaws in Owen's conception of how dinosaurs moved. Instead of the heavy, squat beasts that Owen suggested, further discoveries of fossils revealed that they were more likely lithe and active, with some as likely to move on two feet as on all fours. Intent on depicting New York's dinosaur models "clothed in the forms which science now ventures to define," Hawkins traveled to Philadelphia to seek help from Joseph Leidy, then the country's leading expert on prehistoric life. Leidy let Hawkins view the bipedal *Hadrosaurus foulkii* in the possession of the Philadelphia Academy of Natural Sciences, which when it was discovered in 1858 in suburban New Jersey was the most complete dinosaur skeleton then known. (Owen and Mantell had theorized that what we now know as dinosaurs were prehistoric reptiles from the shape of their teeth; the discovery of a full skeleton seemed to confirm that theory by the presence of a cloaca, so named after the Latin term for sewer, which is the single opening shared by the reproductive, intestinal and urinary tracts in modern-day animals including some reptiles, amphibians and birds.) Hawkins made plaster casts of the thirty-foot-long herbivore's bones and, with Leidy's input, built a metal armature that acted as a form of sinew in steel, allowing him to connect the disparate parts of the animal in their natural positions, creating a hybrid of sculpture and anatomy. The only thing missing was a skull, which he improvised by crafting the head of an *Iguanodon*.

After two months of work, the result was the world's first standing dinosaur skeleton that appeared in a lifelike pose. The specimen was put on display at the Academy in November 1868. Though the museum was only open two days a week, more than 100,000 people—a number more than twice the annual attendance in any year of the institution's history—passed through its doors to gawk at the bones. Their numbers were so great that the Academy's secretary complained that "The crowds lead to many accidents, the sum total of which amounts to a considerable destruction of property, in the way of broken glass, light wood work, &c. Further, the excessive clouds of dust produced by the moving crowds, rest upon the horizontal cases, obscuring from view their contents, while it penetrates others much to the detriment of parts of the collection." Due to the novel problem of what they deemed an "excessive number of visitors," the Academy's trustees decided it was time to charge an admission fee for the first time in the history of the institution, a subtle way to discourage the poor from coming into contact with science.

Hawkins returned to New York to complete the models for the Paleozoic Museum, but soon ran into a problem he had never before encountered while resurrecting the prehistoric world: modern-day politics. The city was run by a corrupt web of politicians headed by William "Boss" Tweed, who suspended funding for the proposed Central Park museum in order to shuffle money to pet projects. When Hawkins complained publicly, a gang of vandals hired by Tweed broke into his studio in the Arsenal building in Central Park on the night of May 3, 1871, and smashed every mold and sculpture Hawkins had built over the prior three years. "Don't you bother so much about dead animals," one of the men reportedly told Hawkins. "There are lots of live animals—you can make models of them." The rubble was dumped in a pit in the northern stretches of Central Park, and the idea of a Paleozoic Museum was buried with it.

The fate of dinosaurs in Manhattan would instead fall to a struggling museum founded by a relatively unknown comparative anatomist named Albert Bickmore. As an undergraduate at Harvard, Bickmore worked in the basement of the university's Museum of Comparative Zoology under Louis Agassiz, a Swiss naturalist who had founded it in 1860 after his popular books and professorship made him one of the country's best-known scientists. A staunch creationist who publicly opposed Darwin and believed that a museum could offer a testament to the variety of God's works, Agassiz nevertheless became famous for his insights into how life changed over time. He was among the first to recognize that the slow movement of glaciers changed the composition of life and rock across Europe, and readily accepted the concept of extinction as the result of catastrophic ice ages—a theory he paired with the argument that God could recreate species as per his heavenly whims.

But it was his views on the systems and categories of life that made Agassiz famous in his own time. He argued that all forms of life—branch, order, species—fell within their own hierarchies, with humans at the top. (This did not apply to all humans, however, as Agassiz claimed to be physically revolted by the first Black person he encountered when he moved to the United States, and argued that non-white humans were of a lesser, distinct species. While he opposed slavery, he did so only because he objected to the "unnatural" intermingling of races.) "I have devoted my whole life to the study of Nature, and yet a single sentence may express all that I have done," he wrote. "I have shown that there is a correspondence between the succession of Fishes in geological times and the different stages of their growth in the egg—that is all. It chanced to be a result that was found to apply to other groups and has led to other conclusions of a like nature."

As a professor, Agassiz was known to be a rampant self-promoter and grating mentor, who confiscated his assistants' laboratory keys if they asked to be paid. Those who could stand his bluster would have to pass a test he called "trials by fish," in which he left students in a room with decomposing specimens and told them to make close observations of the minute stages of decay. "In six weeks you will either become utterly weary of the task, or . . . be so completely fascinated . . . as to wish to devote your whole life to the pursuit of our science," he wrote.

Bickmore passed the test and watched as Agassiz convinced a group of Harvard alumni and the Massachusetts legislature to donate money that allowed the Museum of Comparative Zoology to expand. He soon realized that he would never match his mentor in fame or academic stature. Instead, he asked why he couldn't build a similar museum devoted to exploring the natural world, but at a grander scale. "[W]hen I journeyed for three years in Eastern Asia and over Siberia, I carried with me everywhere two things, a Bible and a sketch plan for a museum in New York," Bickmore later wrote. His reasoning for building a museum dedicated to science in New York City was as practical as his vision was not. "Science does not appear to create wealth directly," he reasoned, so it "must depend on the interest which rich and generous men take in it. . . . New York is our city of the greatest wealth [and therefore] the best location for the future museum of natural history for our whole land."

Bickmore traveled throughout Europe, meeting with members of the scientific elite in London and Berlin and touring the Continent's museums. He returned to New York with a letter of introduction from the head of the British Museum and made his way through the city's aristocracy, pitching the idea of a museum as a way to further the glory of their hometown while preventing rival cities like Boston or Chicago from one day eclipsing it. He eventually convinced nearly twenty of the city's wealthiest men—a group including J. P. Morgan, Theodore Roosevelt (the father of the future president) and A. T. Stewart, who founded the first department store in New York City and amassed a fortune that ranks among the ten largest in U.S. history—to pledge their financial support for a new museum of natural history, and to ask the Board of Commissioners of Central Park to place it on public grounds.

Its name was a reflection of Bickmore's ambition: the American Museum of Natural History. From its birth in 1869, it was meant to be an institution that would have no rival in the country, existing on a higher plane like one of Agassiz's advanced species. Yet Bickmore was not the first person to call something the American Museum. Still alive in memory was a stranger, more cacophonous space that, until a freak fire at the end of the Civil War just four years earlier, was the most popular museum in America—and nearly the exact opposite of the prestigious place of science that Bickmore hoped to build.

Chapter Five

EMPTY
ROOMS

THE MERMAID BURNED FIRST.

A wooden club said to have been used to kill the explorer Captain James Cook in Hawaii fell next in the path of the flames, followed by a case containing live boa constrictors that had dined on fresh rabbits before an audience of schoolchildren earlier that morning. As the fire spread to the upper floors, someone—perhaps a firefighter, perhaps a visitor who panicked and didn't know what else to do—smashed one of the glass sides of an immense water tank with an axe, sending thousands of gallons of seawater cascading down the stairwells and leaving two whales beached on the second floor of a building in Lower Manhattan. Wax figures of Napoleon and Cleopatra, letters signed by George Washington, screeching monkeys of all sizes—seemingly everything that could be conjured by the human imagination soon came tumbling out of the windows and landing in a crowd that had begun to gather on Ann Street.

A firefighter named William McNamara who had visited the building often enough to know its layout ran up the smoky stairwell, past the wax figures of Christ and the disciples and the stage where a "learned seal" named Ned performed twice a day. When he reached the third floor, he kicked open doors until he found

Ann Swan, the eight-foot-tall "giantess" the *New York Times* called an "exceedingly tall and graceful specimen of longitude," and Zuruby, the wooly-haired Circassian beauty, huddling in a corner. He led both women down through the flames to safety, though he refused to go back and grab the $120 in gold coins that Swan said she had hidden in her room. Rumors that a lion had escaped and was prowling around the neighborhood prompted the two women to spend the remainder of the afternoon in the bustling newsroom of the nearby *New York Sun*, where they sipped tea and experienced the novel pleasure of being ignored.

The fire continued to grow, consuming a place that was considered such an essential stop on any New York sightseeing trip that the Prince of Wales had insisted on visiting it when he arrived in the country in 1860. "Birds of rarest plumage, fish of most exquisite tint, animals peculiar to every section, minerals characteristic of every region, and peculiarities of all portions of the earth, costly, beautiful curious and strange, were crowded on the dusty shelves of room after room, where they attracted the earnest attention and studious regard of the scholar and the connoisseur," the *New York Times* lamented on July 14, 1865, as the American Museum lay smoldering on the ground. "Almost in the twinkling of an eye, the dirty, ill-shaped structure, filled with specimens so full of suggestion and of merit, passed from our gaze, and its like cannot soon be seen again . . . for many years the Museum has been a landmark of the city; has afforded us in childhood fullest vision of the wonderful and miraculous; has opened to us the secrets of the earth, and revealed to us the mysteries of the past."

It was a place as famous for its lies as for its truths, a collection of humbugs and spectacles that mirrored the mind of its founder, a man whose genius lay in letting his customers in on the fact that they could not trust him. Phineas Taylor Barnum—P. T. for short—was born thirty-four years after his country's founding, and seemed to embody the uniquely American belief that a good story was more important than fact. This mindset was a gift of his grandfather, who would tell the young P. T. every week that he would one day inherit prime farmland known as Ivy Island, which was destined to make him the wealthiest boy in their hometown of Bethel, Connecticut. When Barnum turned twelve, his grandfather finally took him to tour the property he had been promised. Where he expected to find a place worthy of his dreams, he instead found a hornet-infested swamp. "I saw nothing but a few stunted ivies and straggling trees. The truth flashed upon me. I had been the laughing-stock of the family and neighborhood for years," Barnum later wrote in his autobiography. "My grandfather would go farther, wait longer, work harder and contrive deeper, to carry out a practical joke, than for anything else under heaven. In this one particular, as well as in many others, I am almost sorry to say I am his counterpart," he wrote.

He carried out his first known swindle while working as a clerk at a general store near Bethel, selling thousands of tickets at fifty cents each for what he claimed was a "magnificent lottery." Its winners soon discovered that their prizes were empty bottles and other pieces of junk that Barnum had pulled from the store's unsold inventory. In search of new gimmicks, he tried opening his own store, ran a few more lotteries and briefly served time in jail after founding a newspaper whose specialty seemed to be losing libel suits. He found the success he was looking for in the creation of the American Museum, which opened in Lower Manhattan in the early 1840s.

He was not the first person to open a museum and fill it with the strangest things he could find; he just did it better than anyone else. The first popular museum in the United States was founded by a Philadelphia painter named Charles Willson Peale, who, needing to supplement his income, collected and displayed items ranging from a mastodon bone to Benjamin Franklin's taxidermied angora cat. (That was not quite enough for Peale, who, in a fit of ambition, once asked Franklin if he could display his body, too, once he was done with it.) Known as "cabinets of curiosities," these early museums were often little more than a collection of novel and strange things collected during a wealthy person's lifetime, without any attempt to forge them into a coherent whole or reflection of an idea, and seemed to revel in a lack of order and repel any whiff of seriousness.

Peale initially aimed higher, inscribing the motto "Whoso would learn Wisdom, let him enter here!" above the museum's door, and sprinkled exhibits of taxidermied animals in front of painted backgrounds of their natural surroundings among the less scientific aspects of the collection. "By showing the nest, hollow, or cave, a particular view of the country from which they came, some instances of the habitats may be given," he wrote. Yet poor attendance led his son, Rubens, to persuade his father to lighten the mood by bringing in live entertainers. After Charles retired, Rubens expanded these diversions to include magicians, funhouse mirrors and biological "freaks of nature." Within a year, he doubled the museum's revenue, and he began casting an eye out for ways to expand. He opened Peale's New York Museum in 1825 and established a policy of offering discounts to schools and students. On its four floors, exhibits ranged from colorful gemstones to a calf with two heads to a dog named Romeo who barked answers to questions from the audience. Despite his efforts, the museum could not turn a profit, and Rubens lost it to his creditors in 1830, eventually retiring to his father-in-law's country estate, where he spent the remainder of his life experimenting with mesmerism. (All was not lost for Rubens, however, as a portrait his brother painted of him in 1801 with what was said to be the first geranium ever grown in the New World now hangs in the National Gallery of Art.)

The building and all it contained were eventually purchased by Barnum, who went on to add other exhibits from competing museums and rechristened them all the American Museum.

Outwardly, he promised that the enterprise would be educational, advertising that his collection would serve as an "encyclopedia synopsis of everything worth seeing in this curious world." Privately, he began the search for the sort of spectacle which would bring in the crowds that Rubens had struggled to attract.

The competition was fierce. A German immigrant by the name of Albert Carl Koch toured Manhattan with the mounted bones of what appeared to be an enormous sea serpent nearly one hundred and twenty feet long. He initially dubbed it the *Hydrarchos sillimani* after a friend named Benjamin Silliman. When Silliman objected, Koch went with *Hydrarchos harlandi* instead, after the anatomist Richard Harland, who was not in a position to argue given that he was dead. Only after the specimen was exhibited in Boston and newspapers suggested that it was the remains of a beast that had escaped from Noah's Ark did several Harvard scholars examine it and conclude that it was simply a jumbled arrangement of fossilized whale bones purporting to be a much larger animal. (Koch, by this time, had already taken the exhibit to Europe, where he ended up selling it to a dazzled King Frederick William IV of Prussia.)

Not to be outdone, Barnum purchased the black, shriveled body of what was said to be a FeeJee mermaid that had been caught off the coast of Japan toward the end of the nineteenth century. Three feet long, with sharp teeth, bulbous breasts and scaly skin, the specimen had in fact been masterfully stitched together using parts from a baboon, an orangutan and a shark, among other creatures. Barnum knew that the "all-important question" for audiences was that they be allowed to "see and examine the specimen" themselves to judge its veracity. He exhibited it at the American Museum, where he paid a man claiming to be a scholar to give lectures in the top-floor lecture hall about the natural history of mermaids. When a visitor who had actually been to Fiji stood up and said that there were no such things near that or any island, Barnum's scholar responded that there was no accounting for the ignorance of some men and continued talking. The museum brought in the then-unheard-of sum of $1,000 in revenue the first week the mermaid was on exhibit—three times its normal amount—which cemented its place as the most popular choice of entertainment in a city filled with immigrants who had few of their traditional amusements to fall back on. The American Museum sold more than 30 million tickets over the next decade, including one to president-elect Abraham Lincoln, who walked past a crowd of 250,000 gawking New Yorkers to view the collection on February 19, 1861, while on a tour of the country before his inauguration. After it burned down five years after Lincoln's visit, Barnum lost interest in the business of museums and turned his attention to a traveling circus that one day would reach Topeka, Kansas, and prompt a young Frank Brown to come up with a name for his new baby brother.

Vaudeville theaters and some of the nation's first amusement parks soon filled the space the American Museum had occupied in the country's imagination. Though the name of the institution he envisioned was nearly identical to Barnum's house of fun, Bickmore aimed past the public when it came to the question of how to fund the American Museum of Natural History and focused on a rising class of millionaires looking to cement their place in society through prominent donations. A museum devoted to fusty subjects like rocks and bones should have been a hard sell, especially with the near-simultaneous founding of the Metropolitan Museum of Art, which would eventually sit on the other side of Central Park and promised an eternal association with the beauty and refinement of European masters. Wealthy Americans at the time felt a keen need to purchase art and other cultural treasures as an outward sign of their sophistication, even—and especially—if they were in the relatively rough businesses of railroads or oil. Upon attending a dinner at railroad baron Leland Stanford's San Francisco mansion, one guest remarked that it "looked as if the old

palaces of Europe had been ransacked." Using the spoils of capitalism to bring culture to the masses was seen as a noble calling,
allowing one to act both in ruthless self-interest and for the public
good at the same time. "Think of it, ye millionaires of many mar-
kets, what glory may yet be yours if you only listen to our advice,
to convert pork into porcelain, grain and rice into priceless pot-
tery, the rude ores of commerce into sculpted marble," prominent
attorney Joseph Choate said at a speech at an opening for the Met.

Yet as industrialization remade the U.S. economy and gave the
wealthiest 1 percent of households nearly 25 percent of the country's income, science seemed better equipped to emphasize moral
values of discipline, rationality and the pursuit of knowledge than
art, no matter its prestige. It only helped that the discovery of these
priceless scientific gemstones was often the byproduct of mining
the Earth of its material wealth, producing specimens that would
never be known if not for the thirst of capitalism. In its first years
in a temporary building in Central Park, the American Museum of
Natural History consisted mainly of rare objects of natural origin
that served as centerpieces for exclusive parties. Patrons "prom-
enaded up and down inspecting the numerous cases, and filling
their minds with science, while their ears were filled with the soft
strains of Lanner and Strauss," the *New York Times* noted after an
early gathering of the city's elite.

Few outside the New York aristocracy were interested in or
felt comfortable in such a setting, preferring instead the color and
adventure promised by P. T. Barnum's dime museum. Still, the
American Museum of Natural History doubled down on seriousness, building a massive complex of buildings in Manhattan
Square, a park bounded by Seventy-Seventh Street, Eighty-First
Street, Central Park West and Columbus Avenue, close to the
homes of its benefactors and far from the city's commercial heart.
In 1888, an architect named Josiah Cleveland Cady designed a
massive Romanesque addition in pink granite, featuring so many

castlelike towers, turrets and tourelles that it seemed as if the only thing missing was a moat. Furthering its aim to remain as elite as its benefactors, the museum began funding explorations across the globe, making it something closer to a research institution and, as one trustee put it, "not permit it to be diverted from its original purpose and become a mere show-room of natural curiosities."

Still, it had to do something to bring visitors in the door. When the first section of the imposing addition opened in 1892, the museum struggled to find a way to "sprinkle our wholesome bread with a little sugar," as one trustee put it, and attract a broad audience into a place that by the looks of it was more intent on keeping people out than welcoming them in. Bickmore, fearing that his museum was doomed to become "a stuffed circus, with the chief task of the curators keeping it dusted," began offering free lectures to the city's schoolteachers in hopes that their enthusiasm would spread to their charges.

At the same time, the museum made a more overt appeal to those who felt more at home in P. T. Barnum's dime museum by purchasing a diorama depicting an Arab courier on a camel trying to fend off a Barbary lion that was tearing its claws into the camel's side. Created by Edouard and Jules Verraux, heirs to one of the most famous taxidermy supply houses in France, the work's lifelike poses of both human and animal had made it one of the signature exhibits of the 1867 World's Fair in Paris, with visitors left shaken by the violence of the moment frozen in time. (Only in 2017 did CT scans and an X-ray reveal that the reason the courier looked so realistic was that its head was a human skull covered in plaster, most likely stolen by the Verraux brothers from a graveyard.) Once installed in a prominent place in the American Museum, the diorama shared space with some of the largest fossils in the museum's collection—including a seven-foot-tall Irish elk, a twelve-foot-tall specimen of the flightless giant moa and a mastodon with impressive tusks—all dutifully labeled with their full Latin names, as if to preempt an accusation that anything other than science could account for their position near the front door.

✛ ✛ ✛ ✛

FOR ALL ITS GOOD INTENTIONS, the new American Museum of Natural History barely broke into the stream of life in Manhattan, more of a stately mausoleum tucked away along a wealthy span of the city than anything connected to the here and now. As attendance continued to disappoint, one curator made plain what should be done: the "magnificent museum . . . will need some representation of the giant vertebrate fauna which Marsh and Cope and Leidy have made known to the world." Dinosaurs, in effect, would be the perfect match between the titillating attractions that the public wanted to see and the responsible exhibits that the museum had to offer. The idea was that a visitor, drawn in by the bones of giant beasts, would stick around and see what else could be found in a sprawling palace of science. Marsh, who was not generally interested in what the public wanted, recognized in a speech at the museum that the special appeal of dinosaurs lay not only in the fact that they were so large and strange, but that they spoke to the "great problem" of "the origin of life itself."

The question was how to fill the museum's empty shelves. When he took the job, Osborn pledged that he would turn the museum into a "center for exhibition, publication, and research" in a field "in which America leads the world." He was given $5,000 a year—a sum worth more than $150,000 in today's dollars—to hire assistants and cover operating expenses, a not insignificant outlay at a time when the museum's trustees were beginning to question whether their efforts to build and maintain a little-loved institution had been worth it. The future of the American Museum was effectively in Osborn's hands. Short on staff, money and time, he faced the first true challenge of his life and the first real consequences of failure.

His offers to purchase specimens from both Marsh and Cope

were rebuffed, leaving him in the unfamiliar position of competing in a fair fight with other museums harboring similar ambitions
in cities such as Chicago and Pittsburgh. As he had in Princeton,
Osborn sought to smooth out the wrinkles of his own shortcomings with the experience of others. He hired Jacob L. Wortman,
a moody and virulently anti-Semitic former medical doctor who
had collected for Cope, as his primary assistant. In the newly recognized role of preparator, he hired Adam Hermann, who had
previously worked for Marsh, and gave him the responsibility
of ensuring that fossils would not crumble once excavated from
stone—a job that ensured that the museum would have something
to show for its work in the field. In an essay published near the end
of his career, Hermann noted that in the early days of paleontology
"fossils were dug out of the ground where they were discovered
in the same rough manner that potatoes are dug in the field. They
were picked up in pieces, done up as far as possible in parcels in
the order in which they had been taken up, and then left to the
preparator in the laboratory to fit the fragments together again.
Joining the pieces together, however, was in a great many cases an
utter impossibility, especially if the pieces were quite small and the
fractures not characteristic enough to determine their position."

Over time, Hermann developed techniques, such as the use of
hot glue and plaster, to preserve fossils that are still used to this day,
though a lifetime of frustration with the field paleontologists who
dumped a mess of bones into his laboratory could be sensed in his
gripe that "I have found in my experience that some collectors do
not pay attention enough to the packing and the labeling of the different parcels. This is one of the most important parts of collecting
and should never be neglected." To capture the public imagination
until fossil mounts were ready, Osborn brought in artist Charles
R. Knight to paint murals depicting the prehistoric world, betting
that immersive displays—a nod toward an element of showman-

ship intertwined with paleontology that Marsh despised—would resurrect public interest in the American Museum.

Within two years of taking the job, Osborn had collectors in the American West searching for dinosaur fossils, though they met with little success. Nevertheless, in 1895, he presented a plan to the museum's trustees to open a new Hall of Fossil Reptiles, which he promised would "break down Marsh's work as far as possible" and make the museum "the world's leading center for fossil reptiles."

The only flaw in the plan was that even if the trustees agreed to build a new hall, Osborn still had nothing to display in it. As he prepared for the upcoming prospecting season, Osborn remained desperate to find the bones that would salvage his reputation. He had not only his professional life to consider, but more importantly the opinion of his father. Professorships the elder Osborn could stomach, but not wayward promises of dinosaur bones that if unfulfilled would leave a dark mark on the Osborn name. For all of his willingness to fund his son's life, William Henry Osborn was unrelenting in one aspect: that his son's scientific career "maintain the family prestige."

Chapter Six

A REAL ADVENTURE

HE HAD TO READ IT TWICE TO BE SURE. BARNUM BROWN held the letter from the American Museum of Natural History in his hand, unable to get past the fact that it had come from New York City—a place that may as well have been on the moon, given the fact that he had never been east of the Appalachians, much less seen a skyscraper or walked along a busy city street. Inside, a letter from Wortman offered Brown a spot on the museum's field expedition that summer. If he accepted, he would need to meet up with the party somewhere in the western states by a certain date. Everything beyond that—his destination or the expedition's ultimate aim—was a mystery.

Whereas most people would have a list of questions before committing themselves to a months-long undertaking in conditions that could quickly turn deadly, Brown saw only answered prayers. He immediately withdrew from his classes at the University of Kansas and began cleaning out his room at Professor Williston's home. "Mr. Brown has gone home tonight and Sunday expects to start for Colorado where he will meet Dr. [Wortman] & party of the New York museum and go collecting 'doggie bone fossils' for them," Williston's wife wrote in a letter to her daughter. "It seems pretty cold to start off camping but he thinks they are going to Arizona or New Mexico."

Once he met up with the expedition, Brown learned that he was to start the summer of 1896 in the San Juan Basin in northwestern New Mexico, a parched high desert landscape of blanched rocks and ravines that was once a white sand beach along the shore of a sea covering what is now eastern Utah. As the water receded and tectonic plates shifted, the area became a floodplain where animal carcasses were often covered in sediment before they had the chance to be destroyed by scavengers, beginning the slow process of fossilization. Osborn sent the expedition to the unforgiving terrain in search of the remains of early mammals. Once found, he planned to feature the specimens in an exhibit demonstrating the course of evolution immediately following the disappearance of dinosaurs, putting the museum at the center of the question of why subsequent animal lifeforms never rivaled their immense size.

Dinosaur fossils themselves had proven elusive over the museum's previous expeditions, and Osborn was left with no other plan than hoping that the public would find the bones of prehistoric mammals just as captivating given that they, too, were large. He particularly hoped to find a skull of a *Coryphodon*, a hoofed mammal roughly the size of a cow that was one of the first large mammals to appear after an asteroid hit the Yucatán Peninsula in what is now Mexico about 65 million years ago, killing an estimated 80 percent of all animal life on Earth and wiping away dinosaur species that did not evolve to become what we know now as birds.

Like all expeditions, the crew balanced the competing demands of money and time. Each day in the field that did not result in a museum-quality fossil was a day wasted, and Wortman and his crew lived with the constant fear that Osborn would revoke their funding and spend it elsewhere. Already, his demanding nature and reputation for coldness were well known throughout the small world of paleontology, making the option to work for him the least attractive of most possibilities. Yet the appeal of being associated with a museum based in New York and the unstated promise that

Osborn could secure more funding if needed, given his connections with wealthy trustees, were hard to pass up. If Osborn wanted
a particular fossil, whoever was working for him was to deliver it
or find themselves replaced by someone who would. Though the
museum had few specimens of its own, Osborn refused to put
anything on the exhibition floor that did not meet his exacting
standards. "While it is true that that the collections of the Museum
were unusually rich in the remains of [*Coryphodon*] . . . They are
more or less fragmentary and too imperfect to be used in mounting a complete skeleton," Osborn wrote in a report following the
prospecting season.

The wagon train set off from Colorado in mid-April in weather
that Brown would later remember as "extremely cold and unfavor-
able for rapid travel," and remained in the badlands as the season
turned and the sun scalded them in June. They prospected through
the early summer, trying to imagine the contours of the rivers and
streams of the former world as they searched without success. Each
day fell into the same cycle of blasting, digging and disappointment, repeated under blazing skies and ruminated over each night
under a freezing moon. "After a month's hard work, under the
most trying circumstances, we found ourselves with practically no
results, and what was still more discouraging, with but a few scattered fragments of the animal whose remains we were so anxious
to secure," Wortman wrote.

The youngest member of the team, Brown proved that he
not only had the physical endurance to withstand the grueling
extremes of weather, but the social skills to make it all seem
like a grand adventure. He began trading supplies with a Native
American family who lived near the dig site, offering corn meal
for fresh goat milk—though he stopped drinking it after watching how it was obtained by a local woman. "Sometimes there
were goat droppings in the milk which she skimmed out with
her fingers before sending the milk to us. Having seen the pro-

cess, we lost our appetite . . . but continued to give the Indians meal in exchange for the milk," he later wrote. On other nights, he shared the latest edition of his hometown's newspaper, the *Astonisher and the Paralyzer*, which his parents sent him each week. Wortman, in particular, found the tales of small-town Kansas life enchanting, so far removed from the pressures of working for a high-profile institution in New York. (Wortman would eventually grow embittered with paleontology and quit the field to open a drugstore in Brownsville, Texas, where he spent the rest of his life.)

By late June, the weather was too hot to continue in New Mexico. The expedition had found several mammalian fossils, but no *Coryphodon*. Unwilling to waste an entire prospecting season, Osborn directed Wortman to purchase a wagon and head north. Over the next few weeks, the expedition party trudged seven hundred miles to a quarry in the Bighorn Basin of Wyoming, a region so remote that it might as well have been another planet. Everything necessary for survival—food, guns, bullets and barrels of water—had to be planned and accounted for, given that there were no lifelines to help if the party became stranded. It was as if they were walking off the corner of the map and into the great unknown, all while carrying the expectation that they would bring back evidence of unreal creatures that lived in a world unrecognizable as our own.

The party reached their destination on July 18, and spent the next six weeks collecting in a dry, desolate region nearly devoid of plant life. What it had going for it, however, was a topography that jumped and ebbed, like notes on a sheet of music. A high point more than 11,000 feet above sea level would quickly give way to a valley floor nearly 8,000 feet lower, exposing layers of rock that in some places were more than 2.5 billion years old. Wortman had been the first known paleontologist to explore the region nineteen years earlier, when he traveled north on his own accord while collecting for Cope after learning of the badlands from traders at the remote Fort Washakie.

He later recalled it as "a wild, uninhabited region, save for the occasional visits of roving bands of hostile Indians." In letters to contemporaries who requested his advice before embarking on their own expeditions to the area, he downplayed the threat of violence and revealed that his real fear were the elements. "The exploration of this region is most arduous and difficult," he wrote. "The great scarcity of water in these badland wastes makes it very inconvenient. . . . The broken and mountainous character of the country forbids the use of wagons to such an extent that pack animals are indispensable."

Through hard effort, he learned that what he called blue beds—limestone nodules tinted dark blue by oxidation—were the most likely to bear fossils, and discovered three previously unknown extinct mammalian species in his first season exploring the region. Now, with a complete team from the American Museum supporting him, he once again turned his attention to the blue beds. Over the following weeks, the crew found the remains of extinct horses, monkeys and a wolverine-like carnivorous mammal known as a creodont, but a *Coryphodon* remained elusive—that is, until Brown unearthed a nearly-complete skeleton with a perfect skull, lacking only the hind limbs. Wortman instantly recognized that it would be the crown jewel of the exhibition, delivering exactly what Osborn, still back in his comfortable office in New York, demanded. Having found what they needed, Wortman and the rest of the expedition began their journey back to New York before the weather turned deadly. Brown, however, remained in Laramie through October, spending his time hunting deer and exploring the rich sandstone terrain on the chance that he would find previously unknown fossil beds that would convince the American Museum or one of its competitors to bring him on as a full-time collector.

His ambition was justified: not far away lay the quarry where Arthur Lakes and William Reed, who was then collecting for Marsh, had found the first known *Stegosaurus* in the remote outcrops at Como Bluff. Though Marsh offered only a paltry fifty dollars for the whole quarry, Reed grew to be fanatically loyal to him, going so far as to dynamite fossils that he found but did not plan on taking with him so that they would not fall into Cope's hands. He soon learned, however, that Marsh would not requite his devotion. Fed up with the low pay and Marsh's domineering nature, Reed quit prospecting and became a shepherd. "I regret leaving the Bone business . . . [but] I think it is my duty to look at my own interests first," he wrote.

So began Reed's on-again-off-again relationship to paleontology, leaving him standing somewhere between the respected field collector of an established museum and a man desperate enough to search stones in hopes of finding something of value. A hard winter killed most of his flock, forcing him to look for direction once more. He contacted Marsh, hoping to work for him as a salaried employee the coming summer, but refused Marsh's offer to pay only for the crates of fossils he found interesting. Reed then spent more than a decade bouncing among jobs, ranging from railroad construction to harvesting hay. When, in the early 1890s, the University of Wyoming decided to establish its own collection of vertebrate fossils discovered in the state rather than see professional collectors "take the best things [they know] of and ship them to eastern colleges," Reed was hired at a salary of $1,000 per year. The university soon announced that its "bone room" was as large as the collection at Yale.

While in Wyoming, Brown befriended Reed and learned of potentially rich deposits that had not yet been tapped. He shared this information with Wortman at the American Museum, hoping to prove himself useful enough to be called back for another season of work. "There is a far greater field to be worked up in Jurassic mammals than I had any idea of," Brown wrote. He returned to Lawrence in mid-November, where he jumped back into his coursework as though he had never left. If it was hard to transition from the frontier to the classroom, Brown never showed it, perhaps because he was not a brilliant student to begin with. Still, it was hard not to find him daydreaming of pulling a fossil out of a canyon or gully, far from the flat plains around him. He had tasted the broader world beyond Kansas and found that he measured up, and now he wanted nothing more than to go out and have another chance to prove it.

That spring, Brown received his first letter directly from Osborn, a man he knew only as a mystery. "Dear Mr. Brown, I have had considerable conversation with Dr. Wortman regarding your work during the coming summer and winter," Osborn wrote, in a letter that would forever cleave Brown's life away from the family farm in Carbondale. Though careful not to offer any outright promises, Osborn dangled the lure of New York City before a young man who wanted nothing more than to break free of the boundaries of his childhood. In exchange, he made clear, he wanted to know more about the promising fossil sites that Brown had encountered in Wyoming. "It is possible by prompt application and strong recommendation from Professor Williston that you can secure a [Columbia] University scholarship, which means free tuition, but I do not consider it very probable, there are such enormous demands for these places. . . . I would like a report from you regarding the Jurassic mammal beds, and although I would like you to talk over your plans with Professor Williston, for several reasons I prefer that you should not speak to anyone else about them."

Brown completed the report within a week and sent it to New York, too young and earnest to attempt to conceal his joy. Though Osborn had not formally offered him a job that summer, Brown prepared himself to head back out into the field and hoped that this time he would lead an expedition himself. In a private letter to Osborn, Williston expressed reservations about expecting too much from his student "until [he] has learned where and how to look for the bones," he wrote. Still, Brown charged ahead, his excitement at the possibility of heading back to Wyoming burning so hot that it could have been felt in New York itself. "I can leave here by April 15 or whenever necessary as I have my work nearly completed," he wrote to Osborn. "I am deeply grateful for your kind offer. The University and Museum work is exactly what I desire."

His enthusiasm evident, Brown then detailed his proposal to prospect for the specimens that Osborn craved: dinosaurs. "As to the reptiles," Brown wrote, "I think we can obtain any amount of material. . . . I worked with Mr. Reed a few days in a quarry west of Laramie where the bones were literally packed one on top of the other, nearly all in a good state of preservation."

It was exactly what Osborn wanted to hear.

THE DISCOVERY OF DINOSAUR BONES in the American West that were more gigantic than anything previously imagined took paleontology out of the realm of science and turned it into something closer to a trophy case, reflecting the achievements of those whose money made the discovery possible. Art had long filled this need, allowing a wealthy person to watch their reputation carried upward on a gust of prestige. At the turn of the twentieth century, the forces of industrialization and the rise of corporations handed a handful of men wealth on a scale never before seen. For them, art would no longer do. Something far more rare and difficult would be necessary to reflect the towering position of this new group of gilded tycoons, able to sway millions of lives around the world with their decisions. Nothing less than acquiring and displaying fossils of dinosaurs—the largest and most powerful beasts to ever walk the Earth—could mirror their status in life.

At a time when the churning economy expanded the gulf between the well-off and the rest of the country like never before, the super-rich convinced themselves that funding natural history museums as a form of public education was one way to make inequality into something of a blessing. Steel baron Andrew Carnegie, department store magnate Marshall Field, Wall Street fixture J. P. Morgan: all leaned on grand displays of philanthropy through founding or offering significant support to natural history museums to preempt criticism of their immense wealth. If there was a flaw in the plan, it was that delivering on the potential of the museum often proved harder in practice than in theory. What good would donating hundreds of thousands of dollars to a museum do if no one showed up? "Trustees like to be sure they are getting value for their money," Bickmore, who originated the idea of the American Museum of Natural History, recognized as early as 1873.

Osborn knew that large dinosaur fossils would accomplish that goal, but he had yet to make acquiring a specimen a reality. For the first time in his life, his money and his connections had let him down and opened the door to failure. In Brown, however, he had unexpectedly found a connection to the lucrative bone quarries that had once built Marsh's reputation. Brown was unproven; the fossil beds, however, were not. That calculation allowed Osborn to take a chance he would not normally have committed to and fund a college student still a year away from graduation, knowing that if he succeeded it would change the course of the American Museum. He wrote to Brown offering to back a small prospecting trip that summer, paying him far less than experienced collectors because, as Osborn knew, no matter what price he offered Brown would be a fool to turn it down. Brown jumped at the opportunity, and set off into the wilderness with his future—and that of the museum—resting on his young shoulders.

He returned to Laramie in early May, intending to retrace

the steps of Reed and reopen a quarry which had once proved fruitful. Within days he was in over his head. The roads were a mess; supplies were virtually unobtainable; hiring others to help seemed impossible. Worse yet, the information he had gathered from Reed had proven hopelessly out of date, making him feel as if he had been caught in a lie that would destroy his career before it had even begun. Not knowing what else to do, he wrote a long letter to Osborn, putting his thoughts down on paper while trying to keep at bay the fear that he had somehow screwed it all up.

"I feel pretty badly mixed up in the situation, that I have misinformed you somewhat," he wrote in a rambling letter, describing the sorry state of the quarry he had hoped to reopen. "Marsh had another man continue work there a year after [Reed] left and he went back into the bluff several years until they had a twenty-foot bank to face, now all this has caved in . . . and of course the amount of [dirt] you have to move rapidly increases as you go back," he continued, trailing off in sentences that seemed to have no destination in mind.

He listed the going rate for hiring a man and a team of horses but was too timid to make a decision one way or another. His fear of disappointing Osborn immobilized him, with each pang of doubt reaffirming his insecurity that he was a college kid out of his depth. In time, Brown would become legendary for his cool wit and gregarious nature. Yet at that moment he was still on the border between boyhood and manhood and wanted nothing more than for someone to show him the way. "You see I am completely handicapped," he wrote, displaying a vulnerability toward Osborn that would become a theme of his life. "I don't know how much expense I dare take on myself or what other things might present themselves to your mind. Wire me immediately what to do. I want to do everything to the best advantage of the Museum. If I shall get an outfit will need more money. Meantime I shall gain all information possible and lay over on my expense if necessary."

Whether it was the remote terrain, the farmer's instinct to make a decision that ran through his blood, or the fact that he feared nothing more than being stuck in one place, Brown regained his confidence and wrote another letter to Osborn two days later, spelling out his plan. "You advised me to use my best judgment this summer in regard to all things and I hope I have done right in delaying here until you know exactly how things are," he wrote, his thoughts of self-doubt washed away by forty-eight hours of reflection. "Now I feel confident that it is policy to buy an outfit here. . . . There are certainly other mammal beds there and plenty of reptile material. Shall I not collect everything?" In what surely appealed to Osborn's social and religious codes, Brown's solution was simply to work harder, trusting that his seemingly innate ability to discover fossils would clear any obstacle.

Brown left Laramie in early May for Aurora, a small village fifty miles away whose chief attraction was that it contained a small station on the Union Pacific Railroad, before continuing on to Como Bluffs. Once there, Brown headed to an outcropping dating back approximately 150 million years, when the sagebrush prairie he saw around him was once a humid subtropical plain lush with palmlike cycads and pines that attracted herbivores and the carnivores that preyed upon them. It was during this period, known as the late Jurassic, that dinosaurs began expanding to larger and larger sizes, though paleontologists still do not quite know the catalyst that spurred such a sustained swelling in mass. One theory is that dinosaur bones grew more hollow, like those of birds, making them weigh significantly less than a solid-boned mammal of similar proportions. With this physiological limitation effectively removed, dinosaurs were free to balloon over tens of millions of years before reaching their peak size during the Cretaceous period, a nearly 80-million-year span that began with all of Earth's land mass clustered on just two continents—Laurasia in the north and Gondwana in the south—and ended with the semblance of the continents we recognize today, though India remained adrift in the Indian Ocean and Australia was still conjoined with Antarctica.

Sixty-five million years later, Brown trekked out through the remnants of a once fertile landscape, hoping to uncover the largest bones he could find. With no one to answer to but his own instinct, he spent the first weeks of the summer opening one of Marsh's former mammal quarries. Like the mines they were fashioned on, a bone quarry often consisted of several shafts dug deep into the rock, from which prospectors could search for a seam of fossils. At the end of each season, the shafts would be boarded up and covered as much as possible, a measure intended to protect them from both the elements and the curiosity of others. Brown worked in Marsh's former quarry alone, sifting through the rock and dirt for something that would make Osborn's short yardage of trust in him seem warranted. While working, he could not help but notice an adjacent quarry that had once been tapped by collectors working for Cope. When the Marsh quarry proved fruitless, Brown shifted over and opened up the Cope one. Nearly as soon as he began digging he found dinosaur fossils, so many that he knew there was no way he could excavate them all by himself. The quarry was a "veritable gold mine and I have bones up to my eyes," Brown wrote in a letter to Osborn. Unable to resist seeing it for himself, Osborn left on a rare trip into the field as soon as Brown's message reached him in New York, determined to be on hand when the American Museum secured its first dinosaur.

There, in the badlands of Wyoming, Brown and Osborn met face-to-face for the first time. They had no way of knowing that they would become deeply intertwined in each other's lives, their fates so closely stitched that the success of one would not be possible without the other. Everything about them—their upbringing, their education, their conception of the world and their place within it—were so far apart that it seemed a wonder that the other existed. Yet they were united by their hunger. Osborn, born into privilege and ensured of his natural superiority, would never have come to rely on the youngest son of a poor farmer's family had it not been for the fossils he desperately craved; Brown, a child of the frontier who longed for greater things, would never have tolerated Osborn's outsized ego if not for the chance he offered to leave the farm behind forever. The bond between them would grow over time, yet it would forever be rooted in the fear of failure that each man recognized in the other.

In a photo taken that summer, Osborn and Brown are posed high on a bluff with barren hills behind them. Brown is kneeling, his face tanned by the summer sun, wearing a black turtleneck and a wide-brimmed hat and holding a small pickaxe in his hand. Handsome, strong and composed, he looks as if he belongs, a man equally at ease in a hard landscape or on a city street. At his side, Osborn sits wearing a dark pea coat and gloves, looking as if he has just been plucked from a walk in Central Park. Perhaps to hide his discomfort, he is staring at the camera with an intense glare, like a king reminding his subjects of their place. Between them, sticking out of the rock like a rat's tail, lies the only reason why two life stories so unalike would eventually be bound together: a three-foot-long bone of a *Diplodocus*, which would soon become the first dinosaur specimen in the museum's collection.

Everything about the specimen was big. Its teeth were the size of pencils, each foot larger than a grown man. An adult *Diplodocus longus* stretched up to 92 feet long, making it not only one of the longest sauropods—a group of dinosaurs so named for their graceful, extended necks—but one of the longest animals in the history of Earth. (The longest known dinosaur was likely a North American sauropod known as *Supersaurus*, which may have extended up to 137 feet from head to tail; a modern blue whale, by comparison, tops out at about 105 feet.) Its proportions suggested that nature was attempting an experiment to see just how long an animal's body could sweep: its two-foot-long head was connected to a neck that could stretch more than 21 feet, while its tail spanned another 45 feet behind it. A third of the tail tapered to a thin tip, known as its whiplash, which paleontologists believe the animal cracked to intimidate predators and possibly communicate with other members of its herd. Though immense, *Diplodocus* was strangely light for its size. At 15 tons, it weighed about half that of comparable sauropods and less than a tenth of a blue whale.

With one exposed bone and hopes for a full skeleton beneath it, Brown and Osborn stood at what would be the start of a very complicated puzzle. A full dinosaur specimen was invaluable, its unmistakable form a guarantee that crowds would come to any institution where it was displayed. A random leg or arm bone or a scattering of teeth, meanwhile, would be of intense interest to paleontologists but make no ripples on a museum floor, uninteresting and unremarked upon by anyone who happened to walk by. Though he hoped to supervise the process of excavation to ensure that no harm would come to the museum's new prize, Osborn realized within a few days that the process would take weeks, far more time on the frontier than he could stand. Before he left, he sent messages to Wortman, who was then prospecting in Colorado, and a field crew working in Nebraska to abandon their projects and rush to Wyoming. Not long after he arrived, Wortman discovered another giant fossil while attempting to excavate the *Diplodocus*. The new specimen was an *Apatosaurus*, which when alive had stretched 75 feet long and stood on four massive, pillar-like legs.

The work continued through the heat of August. Wortman and the crew of reinforcements worked on the *Apatosaurus*, which Brown described in a letter to Osborn as "beautiful bones, as perfect as any I have ever seen from the Jurassic." Brown took on the *Diplodocus* alone, covering the top of each portion of the skeleton he uncovered in a jacket of plaster of Paris before digging a small trench underneath it and wrapping it again, enclosing it in a white cocoon that looked like a misshapen beehive. For the bones too heavy or impacted in rock to excavate from above, he grabbed pieces of timber and built makeshift mines to attack them from below. Dozens of crates filled with wrapped bones were loaded onto trains heading east as the prospecting season wound down in September, the freight cost of shipping tons of fossils paid for by a connection of Osborn's father.

With their first dinosaur specimens in hand, Osborn and Brown accelerated the trajectories of their lives. Osborn, never one to play down his successes, published a report in which he inserted the fact that he had been the co-discoverer of the two specimens uncovered that summer. "So far as the skeletons themselves are concerned . . . They are, perhaps, by far, the most complete and perfect of their kind that have ever been collected and will make magnificent material for purposes of exhibition," he wrote. He sketched out plans for two freestanding displays that would mount the creatures in lifelike poses, a feat that had never before been attempted with fossils so large, given the amount of weight that any support would need to hold. He had long harbored big dreams for the American Museum; now he owned the bones that could make them possible. The dinosaurs would be not only a validation of his career, but a monument to his genius.

For Brown, meanwhile, the adventure was only beginning. Toward the end of the summer, Osborn informed him in a terse letter that he had been selected for the full scholarship to Columbia University. He had braved the wilderness. Now his prize was waiting for him: New York City.

Chapter Seven

FINDING A PLACE
IN THE WORLD

IN THE LATE SUMMER OF 1897, BARNUM BROWN STEPPED off a train from Kansas into one of the busiest stations in the world. Men clad in bowler hats and ladies in prim Victorian dresses buzzed about, set for destinations across the country. Never before had the world felt within such easy reach. With the right ticket in his hand, Brown could be on a coast three thousand miles away within less than eighty hours, flung across a continent on the power of steel tracks. A few months earlier, he had been excavating the remains of the prehistoric world; now, as he made his way to a steam ferry docked along the humid New Jersey riverfront and glimpsed the forest of skyscrapers on the other side of the Hudson, he saw for the first time the city that felt like the future.

New York was a place of extremes, filled with a population unencumbered by history and anxious to start anew. A few months after Brown arrived in the city, more than ten thousand people in Union Square braved a bitter New Year's Eve rain to watch a parade of electric floats celebrating the Festival of Connection, honoring the merger that went into effect at midnight joining Manhattan, Brooklyn, Queens, the Bronx and Staten Island into the modern City of New York. The consolidation of power joined the nation's first- and fourth-largest cities into one super metropolis, a giant with a footprint stretching nearly 303 square miles. The rush toward bigger and stronger things left some worrying that the provincial character of Brooklyn would be corrupted by the rise of corporations, fueled through a bold new era of securitization on Wall Street that created institutions whose power and size was unimaginable just a few decades before. "Brooklyn has repeatedly shown herself to be the most independent urban community in the world," St. Clair McKelway, the editor of the *Brooklyn Daily Eagle*, told a considerably more somber audience gathered in mourning that New Year's Eve on the steps of Brooklyn's City Hall, soon to be known as Borough Hall. "There need be no apology for the poverty of Brooklyn. It is an honorable poverty."

Everywhere Brown looked were signs of the modern era overtaking the past. Directly ahead in Lower Manhattan loomed the Gothic spire of Trinity Church, which at 281 feet had stood as Manhattan's tallest structure for the past fifty years and was the traditional site of the city's New Year's Eve celebrations. Its reign would soon end, however. A few blocks away workers were clearing Revolutionary War-era stables and squat general stores for the construction of the Park Row Building, whose soaring twin copper-tipped domes would top out at 391 feet and make it the tallest building in the world when it was completed in 1899. Architecture critics were not sure what to make of this new behemoth, one of the first buildings to properly be called a skyscraper. "New York is the only city in which such a monster would be allowed to rear itself," one paper argued, while another deemed it a "horned monster." At the lip of New York Harbor, Brown could spy bricklayers working amid the scorched ruins of the immigration station on Ellis Island, which had been destroyed by a fire just a few months earlier and was already in the process of being expanded and rebuilt. Far uptown, within a few minutes' walk of his destination, Columbia University, stood the newly-built Greek mausoleum holding the tomb of President Ulysses S. Grant, who remained one of the most popular figures in the nation for his role commanding the Union Army in the Civil War. More than sixty thousand soldiers from throughout the country had marched in a parade before Grant's Tomb earlier that spring, as if intent on burying not only the war hero but the scars of history. "Never before in the history of parades were there fewer accidents or did the police have less trouble in keeping the throng within the lines," the *New York Times* reported the following morning.

Brown was one of millions of new arrivals to the city, many of them hailing, like him, from small or rural towns and driven by the chance at a better life. Over 1.5 million immigrants had passed through Ellis Island over the previous ten years, many of them coming from poor provinces in Germany, Ireland and Italy, willing to brave the unknown. Nearly 100,000 African Americans from the South migrated to New York over the same period, finding a foothold of freedom within memory of the abolition of slavery.

Those riding the swelling fortunes of Wall Street made their homes in grand new apartment buildings that stood as shrines to the new era of money. At the recently-opened Dakota Building on Central Park West, residents enjoyed the marvels of central heating and an in-house electric power plant. The building proved so popular that it led to a boom in luxury housing, filling what had been considered a remote part of Manhattan with opulence. While the rich moved higher into the sky, more than 2.3 million people—two-thirds of the city's population—crammed into squalid tenements downtown, barely able to survive. Rooms overflowed with people; children went hungry; whatever dollars were on hand never seemed to be enough. "Once a nurse navigated the hallways and reached her destination," a medical student named Lillian Wald wrote in an essay describing her experience in the tenements of the Lower East Side, "it was not uncommon to go in the daytime into the closet-room with candle or lamp to be able to see the patient at all; not uncommon to go into the house and see ten or eleven people occupying two small rooms—people who have been working all day, for the night's *rest* stretched on the floor, one next to the other, dividing the pillows, different sexes not always of the same family, for here are 'boarders,' who pay a small sum for shelter among their own, the family glad of this help to the rent."

The poverty of the growing city, especially among its youngest residents, pushed social reformers to call for playgrounds, parks, schools—anything that would offer a citizen of the slums a glimmer of life beyond their narrow horizons. "With no steady hand to guide him, [a boy from the tenements] takes naturally to idle ways . . . the result is the rough young savage, familiar from the street," wrote Jacob Riis, a Danish immigrant whose work reporting on New York's slums led to his landmark book *How the Other Half Lives*. "Rough as he is, if anyone doubt that this child of common clay has in him the instinct of beauty, of love for the ideal of which his life has no embodiment, let him put the matter to the test. Let him take into a tenement block a handful of flowers from the fields and watch the brightened faces, the sudden abandonment of play and fight that go ever hand in hand where there is no elbow-room, the wild entreaty for 'posies,' the eager love with which the little messengers of peace are shielded, once possessed; then let him change his mind."

The museum into which Brown would soon walk for the first time was one of the few places in a position to unite the far reaches of the city's rich and poor, binding the population together through a shared fascination with the range and beauty of the natural world. It had yet to make much of its opportunity, however. Its grand experiment that science could be a common ground for the sprawling metropolis remained unproven, its halls more likely to hold academics or wealthy students than those who would have the most to gain from even a brief exposure to a place devoted to learning. In its annual report that year, the museum revealed that it required a steep drawdown from its endowment and several charitable gifts to meet its operating deficit. Unless it could find a way to bring in more visitors to make up for its small amount of public funding, the report warned, the institution might never be profitable, leaving it forever at the whim of Wall Street and unreliable donors. "It will be seen that the amount received from the city is not sufficient to maintain the Museum," the annual report cautioned.

Nothing was working. The Barbary lion diorama led to no noticeable uptick in visitors; the mammoth exhibit and giant elk were not talked about in the city's poorer neighborhoods in the same tones of awe that had once accompanied the latest exhibit at P. T. Barnum's famous and now lost museum. Despite all of its efforts, the museum staff had accomplished the impossible: making a collection of some of the rarest and most fascinating items ever collected seem boring and unworthy of a person's time.

Its newest attractions that year were castoffs and giveaways, a collection of curios unloaded from the closets of its wealthy bene-factors and amateur collectors. The annual list of additions seemed more like the inventory of an estate sale than anything designed to get the public clamoring to get into the building. James A. Bai-ley, who founded what became the Ringling Bros. and Barnum & Bailey Circus, donated the skeletons of two camels, two kangaroos and an Indian elephant. An insect enthusiast by the name of Henry C. Pratt presented a gift of thousands of dead termites and ants collected in Haiti. And a Miss Annie Peniston sent a shipment of shells she had gathered in Bermuda. In a move of desperation, the museum obtained the bodies of animals and fish that died in the city's zoos and aquarium and stuffed them for display, as if the pub-lic would be more interested in seeing taxidermied corpses than they would have been watching those same creatures swimming and breathing just a few months before.

Though the museum had the skeletons of the *Diplodocus* and the *Apatosaurus* unearthed in Wyoming sitting in storage, it would take months if not years to prepare them for display. Everything about paleontology—the trial and error involved in finding a fossil, the work to pull it out of the ground, the long lag between the discovery of a specimen and the time it took to clean the fossils and mount them on a frame—was slow at a time when the museum needed a hit right away, leaving it to improvise and make do until it could reveal its treasure. With no dinosaur fossils ready to show, Osborn filled the halls of the Vertebrate Paleontology wing with large charts showing the succession of animals found in rock layers in North America and watercolor paintings of untouched landscapes. His main attraction was the skeleton of a three-toed rhinoceros, mounted so that only an astute viewer would notice the two beams keeping the specimen upright. He knew that it was not enough, though he maintained an outward face of optimism. "In the Department of Vertebrate Paleontology we have continued to devote the greatest care and study toward arousing the interest of the public in our exhibits," he wrote.

As Brown walked through the museum's halls for the first time, weaving among the construction zones of the five interlocked buildings that when finished would make up the facade along Seventy-Seventh Street, he passed through empty rooms earmarked to hold bones that he had not yet found. He knew, more than anyone, how an exposure to science at a young age could change the course of a life, and he felt the weight of finding attractions that could hold the interest of children whose parents had, like him, traded a rural life for one that held greater promise. He had already traveled farther from Carbondale and its routines of selling coal and chasing chickens than he had dared to imagine, yet it only made him realize that the world was bigger than his childhood dreams. Like countless other young men and women enchanted by their first exposure to New York, the daily toil of life in the city only increased his ambition.

He began a new routine, living in a room of a house on East Sixty-Seventh Street owned by a doctor who had his medical practice on the first floor, working at the museum as an assistant curator of vertebrate paleontology and taking graduate courses at Columbia, though he had not yet completed the coursework to earn his bachelor's from the University of Kansas. He was one of the few students at the university who did not come from money; he was there for the education and the increase in rank it would signal rather than for the connections he might make. The stiff, slow world of formal education did not excite him and he soon fell behind, daydreaming about another dig rather than committing to the slog of completing another paper.

In the field, he would have outshone every other student, able to put into practice what they could only discuss in theory. Yet the world he was attempting to step into was not the frontier, and the skills that helped so much there—a willingness to try anything, to bend the rules and to rely on physical strength and endurance to outlast any competitors—were of no use in a lecture hall. He could barely tolerate going through the motions of academia when he longed for the adventure of discovering the fossils that would form the basis of ideas his fellow classmates were now discussing. He belonged to the paleontology of dirt and rocks and dust, and instead was asked to make do in a world of paper and books and pencils.

The classroom could also not compete with the appeal of a beautiful blond woman who lived on another floor of his boarding house. Courting women came naturally to him, leaving fellow members of his field expeditions to joke that he could find a lover even in the most barren places on Earth. Those affairs were often short-lived, done in as much by the ephemeral nature of a dig as by Brown's need to pursue something else that caught his attention.

But in New York, the city that offered a young, attractive man more options for pleasure than any other, he for the first time encountered a person who appealed to his brain as well as his body.

Like Brown, Marion Raymond was a part-time graduate student at Columbia, finishing her master's degree while teaching high school biology. But that was about as far as the similarities went. She was the daughter of a distinguished lawyer and educator from the tidy hamlet of Oxford, New York, a woman more comfortable with a book or in a laboratory than in one of Brown's all-night poker games; he was a farmer's son who despite his brilliance and long career would author but two scientific papers and could never abide the idea of staying in one place for long. Yet something about the two clicked, each finding in the other what they had been missing in themselves. As Barnum and Marion began spending long hours in each other's worlds, he experienced the novel sensation of slowing down and conversing with a woman whose mind equaled his own, while she grew comfortable with the pleasant chaos that was Barnum's drive to experience all that the world had to offer.

For the first time in his life, Barnum began to feel the restlessness which had been a part of him since childhood seep away. Before Marion, before New York, before the museum, he had always felt the need to chase something down; now, he felt the novelty of contentment, sprouting up from soil that he did not know he had within him. It was all so early, and so quick. He had been in New York a little over a year, and in that time had braved a classroom filled with men whose privilege far outshone his own, and held down a job that was his childhood fantasy while living in a city as different from the farm back home as he could imagine. And now he felt himself falling in love for the first time. For a man who could never feel peace, his world was finally settling into place.

And that was when Osborn came into his office one morning and told him to pack his bags for Patagonia.

Chapter Eight

The Uttermost Part
of the Earth

The Patagonia region of South America sits at the bottom of the world, extending over 900,000 square miles through Argentina and Chile. Sandwiched between two oceans, a strait and a river, its interior offers a survey of the Earth's extremes: mountainous peaks where sudden violent snowstorms appear without warning give way to nearly barren deserts marked by deep gorges, while lush forests taper into the largest ice fields in the Southern Hemisphere outside of Antarctica. The region culminates in an archipelago that Spanish explorers named Tierra del Fuego, or Land of Fire. There, hundred-foot waves routinely batter boats trying to round Cape Horn, leaving its shores a graveyard of ships and sailors alike.

"The calms here are unlike those in most parts of the world," wrote Richard Henry Dana Jr. in *Two Years Before the Mast*, his 1840 recollection of sailing on a small clipper ship around the Cape from Boston en route to the still-Spanish colony of California. The ship took four days to pass the Cape, battling hail, snowstorms and wind so severe that it threatened to pull a sailor off his feet. "Here there is generally so high a sea running, with periods of calm so short that it has no time to go down, and vessels, being under no command of sails or rudder, lie like logs upon the water."

A few years before Dana's voyage, a 90-foot-long ship called the HMS *Beagle* had appeared off the Cape while on a five-year-long voyage around the world. Among the sixty-eight men on board was a twenty-four-year-old son of a society doctor named Charles Darwin. He paid his own way to serve as the voyage's naturalist, intent on collecting and cataloging plants and animals unseen by English eyes. In truth, he was mainly there to keep the twenty-six-year-old captain, a fellow aristocrat named Robert Fitzroy, from getting lonely. (Fitzroy took command of the *Beagle* after its previous captain, Pringle Stokes, grew increasingly despondent and isolated while on a mission to survey the coast of Patagonia, a region which he described in his journal as consisting of such dreary and brutal weather that "the soul of man dies in him"; Stokes locked his cabin and shot himself in the head two years into the voyage, dying twelve days later.)

Darwin was decades away from formulating the theory of evolution by natural selection that would make him a pillar of British science, held in such high esteem that upon his death he would be buried in Westminster Abbey alongside kings. Instead, he was a wealthy young man who had turned to natural science after an attempt to follow his father into medicine ended when he threw up after watching his first autopsy, and he jumped at the chance to travel the world because he had little else to do. He was expected to dine with Fitzroy and serve as a trusted counsel he could talk with, but beyond that to mainly stay out of the way. The sense that Darwin was a late addition to the voyage was underscored by the fact that, with no other room on the crowded ship, he slept in a hammock slung over a drafting table. He spent the entire five-year journey battling violent seasickness, heaving with each passing swell. "I hate every wave of the ocean, with a fervor, which you, who have only seen the green waters of the shore, can never understand," he wrote in a letter to a cousin.

When he was not seasick, Darwin drew in the richness and

variety of ocean life he had never before encountered. He noticed the subtle artistry in oyster shells embedded in rocks and the gracefulness of tiny fish and plankton he pulled up in his net, all the while wondering why God spent time perfecting something so few people would see. "Many of these creatures so low in the scale of nature are most exquisite in their forms & rich colours," he wrote. "It creates a feeling of wonder that so much beauty should be apparently created for such little purpose."

Each day he encountered another mystery of nature, and, with nothing else competing for his time, he plunged headlong into trying to figure it out. His enthusiasms shifted with the changing landscape. When at sea, he formulated theories as to how coral reefs formed; when on land, he wanted to collect every animal in the jungle. "Here I first saw a Tropical forest in all its sublime grandeur. — Nothing, but the reality can give any idea, how wonderful, how magnificent the scene is . . . I am at present red-hot with Spiders, they are very interesting, & if I am not mistaken, I have already taken some new genera," he wrote.

After nearly a year at sea, the *Beagle* reached Patagonia. The solitude of the landscape unnerved Darwin, and he spent his hours on watch painfully aware of how far away he was from home. "The consciousness rushes on the mind in how remote a corner of the globe you are then in; all tends to this end, the quiet of the night is only interrupted by the heavy breathing of the men & the cry of the night birds," he wrote. His first encounters with the Yamana, the tribal people who lived in the region, left him with the sense—common among Victorians at the time—that he was glimpsing a lesser form of human. (Indeed, on the same voyage, Fitzroy returned home with a man named Orundellico, a member of the Yaghan Indigenous group whom Fitzroy had kidnapped a year before on the beach and brought to England as proof that he had reached the continent. Orundellico, whom the English called Jemmy Button, would meet up again with Darwin a year later, having shed his short-cropped hairstyle and English clothing and clad in nothing but a blanket around his thin waist. "We had left him plump, fat, clean, and well-dressed; —I never saw so complete and grievous a change," Darwin wrote.)

Drawn by a false sense that he was seeing an untouched vision from the romantic past, Darwin was both fascinated and repelled by the humans he met. "I do not think any spectacle can be more interesting, than the first sight of Man in his primitive wildness," he wrote. "It is an interest, which cannot well be imagined, until it is experienced. I shall never forget, when entering Good Success Bay, the yell with which a party received us. They were seated on a rocky point, surrounded by the dark forest of beech; as they threw their arms wildly round their heads & their long hair streaming they seemed the troubled spirits of another world."

Captain Fitzroy decided to spend the winter in Patagonia, giving Darwin a chance to explore the rivers leading up to the Andes. There, he collected samples of everything he could find, stacking thistle next to deer skins next to shells, and lamenting in one letter the strain it took to capture some of the rare beetles he came across. The region is "most singularly unfavourable to the insect world," he complained. Yet he also began to realize that the undeveloped region was in fact a giant graveyard, filled with the bones of enormous animals that no longer inhabited the Earth. Patagonia was "a perfect catacomb for monsters of extinct races," he marveled. He found huge prehistoric armadillos and giant birds, all the while wondering why some lifeforms had apparently shrunk over geological time. "It is impossible to reflect on the changed state of the American continent without the deepest astonishment," he wrote. "Formerly it must have swarmed with great monsters: now we find mere pygmies. . . . What has exterminated so many species?"

Darwin would ponder how and why some lifeforms survived and others did not for the next twenty years. He experimented with seeds in seawater, satisfying his hunch that they could withstand the lurching sea until they washed up on a far island, and tested how genetic variations spread through colonies of pigeons and bees. Finally, eight years after the death of his oldest daughter from typhoid erased the last vestiges of his Christian faith, he published *On the Origin of Species*, an explanation of evolution driven by natural selection that suddenly made everything from the colors of a peacock's feathers to the colossal bones uncovered in Patagonia make sense.

Among scientists, the region would have remained best associated with Darwin's expedition had it not been for an Argentine paleontologist by the name of Florentino Ameghino. The son of Italian immigrants who were unable to afford him a formal education, Ameghino first discovered natural history through books, and then through long walks in his adopted country. His genius was apparent despite his poverty and he eventually became a professor of geology and mineralogy at the University of La Plata, where he found himself embroiled in the most burning question of the era: if Darwin's theory of evolution through natural selection was indeed correct, then how and where did humans and other mammals originate? While European and American scientists centered their view on Africa, Ameghino argued in a well-publicized paper that Argentina was the cradle of human life. Ameghino's theory drew fierce criticism from European scientists who felt that it lacked rigor, but others found it credible enough that expedition teams from around the world set out to Argentina to investigate whether this little-known professor was on to something. Cope, for his part, called the paper "a monumental work, such as can only be produced under circumstances which seldom concur."

Osborn lived for any chance to make the American Museum part of an intellectual debate, searching for any method to further his own reputation. Ameghino's theory proved especially difficult for him to pass up. Osborn had long disbelieved the prevailing notion that humankind originated from a common ancestor in

Africa. Over time, his suspicion would harden into something altogether more cruel, infused with the vile notion that fixed racial traits in humans meant that light-skinned Europeans were naturally smarter and stronger than dark-skinned Africans, whom he considered less evolved. In one particularly shameful incident in his life, Osborn served on the board of trustees of the Bronx Zoo when it installed a twenty-three-year-old Congolese man named Ota Benga in an exhibit in the Monkey House, and refused to meet with Black clergymen who, appalled at its implications, protested that "We think we are worthy of being considered human beings, with souls." Osborn's racism was common among his peers. While Black Americans protested the spectacle, a medical doctor by the name of M. S. Gabriel wrote an essay for the *New York Times* in which he declared, "I saw the pigmy on exhibition, and must frankly state that the storm of indignation which some well-meaning clergymen are trying to raise around it is absurd. The unprejudiced observer cannot possibly get the impression that there is in the exhibition any element implying the slightest reflection upon human nature or the colored race."

Osborn, however, took this a step further. Rather than viewing skin tone as just one of many adaptations that evolve in a species over generations, he held that racial attributes were a sign of parallel evolution, suggesting that light-skinned humans did not evolve from apes but from some other species yet to be discovered. Throughout his life, Osborn would jump at any evidence that appeared to justify his pet theory, which went unchampioned by his peers, and never wavered in his refusal to accept that a common thread of humanity lay below the veneer of skin color. His racism would infect the American Museum of Natural History, leading to displays and exhibitions that implied that white Protestants of what Osborn called the Nordic race were the pinnacle of billions of years of evolution.

As those theories hardened in his head, he struck a deal to send a museum staffer along on a trip organized by his college friend William Berryman Scott, to search for clues to mammalian evolution in Patagonia. Osborn chose Brown, perhaps struck by his outsized role in obtaining the museum's only dinosaur specimens. If Brown could find a *Diplodocus* alone, Osborn's thinking went, then who knows what he would bring back from a region known for its abundant fossils. The decision infuriated Jacob Wortman, who had lobbied hard for the opportunity. Passed over for a man barely out of college who had served as his trainee not long before, Wortman began nursing a grudge against Brown that he held on to for the rest of his life.

BROWN KNEW NOTHING OF OSBORN'S plans and little of Patagonia's history until the morning of December 7, 1898, when he arrived at the American Museum shortly before nine in the morning after trudging through snow piled high along the city's streets. "Before I had taken my hat off, Professor Osborn called me into his office," Brown later wrote, describing the two-minute conversation that upended his life. Osborn asked him, "Brown, I want you to go to Patagonia today with the Princeton expedition. . . . The boat leaves at eleven; will you go?" Without hesitation, Brown replied, "This is short notice, Professor Osborn, but I'll be on that boat."

He rushed home to pack a small bag and raced back through the snow to Pier 45 in the West Village. There he boarded a Grace Line freighter named the *Capac* bound on a nonstop voyage of over six thousand miles to the port of Punta Arenas on the Strait of Magellan. As the ship left New York harbor and the skyscrapers of Manhattan slowly sank over the horizon, Brown found himself yet again in an alien environment. Unlike the frontier, however, his body was not equipped for this. It was Brown's first time at sea, and he struggled to adjust to life aboard ship. He "soon was a victim of seasickness, first hoping I would die, and then afraid I wouldn't," he later wrote. For thirty days, the ship steadily moved south. Christmas and New Year's Day passed by, two days spent like any other staring out at the featureless ocean. The ship spotted land just once over the four weeks as it sailed by Pernambuco on the coast of Brazil.

To pass the time, Brown played poker with the expedition's leader, John Bell Hatcher, who was equally famous in the field of paleontology for his part in discovering *Triceratops* and for his fallout with Marsh, who refused to let Hatcher name the species because he had paid for the expedition. Fearing that Osborn would prove to be another tyrant in the same mold, Hatcher had not wanted Brown or any other member of the American Museum on the trip to Patagonia, but was overruled by Scott. If he couldn't get rid of Brown, he could at least empty the man's wallet. Hatcher spent every day and night over the four weeks it took to get to Patagonia systematically draining Brown of every cent to his name, one poker chip at a time. " 'Brown,' he would say, 'I hate to take your money as I know your salary is only fifty dollars a month,' " Brown later wrote. Only one fluke hand kept Brown from destitution. As they neared land, Brown somehow won back almost everything he had lost over the previous month at once, saved by the kind of outlandish bet that would become a constant of his life.

The *Capac* finally reached Punta Arenas in early January, and its crew rushed to gather provisions for the expedition ahead, like squirrels surprised by an early snowfall. Brown walked along the streets of the busy port city, taking in its mix of English, German and Latin merchants as he tried to make himself useful. The farm boy within him couldn't help but notice the incongruity of advertisements for high-end Italian and Spanish wines posted next to the offices of sheep-dealers, the economic lifeblood of the region. Though he did not understand the languages flittering through the streets, the city felt strangely familiar, the sort of hive of activity on the border of an empty expanse that he had trudged through with his father as a child and would encounter again on frontiers throughout the world. Hatcher directed his men to stock up on enough food, shovels, pickaxes and whiskey to last for several months. He acquired two teams of horses and harnessed them to a cloth-covered Studebaker wagon. This would be the first time that a scientific expedition attempted to cross the vast, windswept Patagonian pampas on a set of wheels. (The Studebaker brothers would introduce their first electric car four years later, and would continue producing canvas-covered wagons alongside automobiles until the 1920s.)

Hatcher jumped on the back of a horse and rode ahead, alone, to check on the condition of a fossil bed he had uncovered the previous spring, leaving Brown and another assistant to follow behind on a four-hundred-mile route along smooth, flat stones that had once been covered by the receding ocean. Neither knew the path beyond Hatcher's directions, and took to a system in which Brown drove the wagon while his partner scouted the trail. A few days into the trip, Brown was holding the reins when the ground around them began to recede and fall, as if a plug had been pulled from below. He yanked the horses as hard as he could to the side and tried to race ahead, not knowing whether the land itself was crumbling. When he looked back, he realized that the wagon had barely escaped falling into a deep, quicksand-like bog of mud known as a soap hole. "Had the team and wagon gone into this mass, we would have all gone down without anyone knowing where we had disappeared," Brown wrote, spooked at the realization that he was crossing a place where even the ground seemed out to get him.

They reached the foothills of the Andes and began the slow trek upward. Once the trail became impassable, they ditched their wagon on a dry lakebed and continued on horseback. Finally, in March, Brown reunited with Hatcher. The expedition team was almost immediately caught in a blizzard, forcing them to release their horses to seek shelter. Brown, Hatcher and the others crammed onto one single bed in a tent and remained there for two days to preserve body heat. The storm eventually passed, allowing a brief window of time to dig before another sudden violent snowstorm again whipped through, halting everything once more.

Hatcher's calm demeanor while battling the weather made an impression on Brown, who at the age of twenty-five was still puzzling out the question of who exactly he was going to be as he stepped further into adulthood. There was enough alike in Hatcher's background—a childhood on a farm in the Midwest, a brief stint in a coal mine, college in his home state of Iowa—for Brown to see a reflection of himself and a model of what he could become. While Osborn's wealth and status would forever make him a man apart, Hatcher was someone whom Brown found himself looking up to as a mentor. He admired how Hatcher carried himself with a sense of purpose; furthermore, he had demonstrated an ability, which Brown had not yet mastered, to mesh the scientific world of papers and lectures with the real-life skill to acquire the fossils that would push the field further. Through his physical abilities and success in Wyoming, Brown had already shown his worth outside of the classroom, yet he kindled doubts whether his academic skills would ever catch up. Hatcher was "a truly remarkable man, with few vices and more virtues than are found in most men . . . as a worker he was indefatigable," Brown would later write. "He would ride off alone in an uncharted area, with only his blankets, revolver, and a pocket full of salt, living off the game of the land as he travelled. His geological observations were, to my knowledge, accurate; and as a collector, no one ever surpassed him."

After two months and several snowstorms, Hatcher had to admit that the trip was proving to be a bust. The fossil beds he had expected to find in abundance were nearly nonexistent. After a final effort to trace a seam of fossils proved fruitless, Hatcher announced that the expedition would return to Punta Arenas and head home. They descended from the mountains and made their way back to the city, their empty wagon reminding them of their failures with each easy rotation of its wheels. While Hatcher consulted with travel agents for the next ocean liner to New York, Brown explored nearby marine beds, unable to stand the thought of disappointing Osborn so thoroughly. By chance, he found the skull, jaw and vertebrae of a toothed whale that had lived in the Miocene, some five million years ago.

It was enough to keep him from getting on the boat home. Never before had Brown headed off into the field and come home empty-handed, and he did not want to start on a trip where he carried the weight of Osborn's expectations on his back. As Hatcher and the other members of the expedition boarded the ship, Brown was far away, gathering supplies for what would become a six-month mission prospecting alone along the coast of South America. With no one to answer to, he allowed himself to get lost in the natural world, his ambition and curiosity merging to form a single-mindedness which bordered on the reckless and pushed him to take risks that older, more experienced prospectors would never dare.

At Cañon de las Vacas, he built a sling out of a tarp and used it to hang off a cliff face so that he could chisel out what turned out to be an armadillo-like creature known as a *Propaleohoplophorus*, a specimen that now stands in the American Museum's Hall of Mammals and Their Extinct Relatives. Near the outlet where the Gallegos River spills into the Atlantic, he dodged incoming waves and the occasional shark to dig up a fossil he thought he saw exposed at low tide. After several hours of work, he pulled out the skull and jaws of an extinct creature known as an *Astrapotherium*, which looked like a cross between an elephant and a hippopotamus yet was unrelated to either. Not long after, Brown found himself standing chest-deep in the ocean while locked in battle with an octopus that refused to get into his pickling jar, too immersed in his task to realize that the tide was coming in. Only when his horse whinnied did Brown realize that the water was up to its belly. He grabbed the octopus and jumped on the horse's back and rode it to shore, trying to avoid the deep holes hidden by the waves.

A few weeks later, he came across sixty slaughtered lambs while riding alone through rocky foothills. A survey of the area uncovered tracks of a mountain lion and its cubs, which he followed to the mouth of a cave. He lit a candle, drew his revolver and walked in. "I had proceeded a short distance when all at once there came a roar—apparently back of me," he later wrote. "Holding the candle up, I wheeled, and saw the reflection of two eyes. I fired between them . . . I heard a kicking; then all was still . . . I think this terrifying experience was when I lost my hair—figuratively scared off."

By October, Brown had amassed an impressive haul of fossils. "This collection includes a more varied fauna than that obtained by Mr. Hatcher from Gallegos . . . this will make a great exhibit," he wrote in a note accompanying the shipment to New York, his growing confidence apparent on the page. Over six months of solitude, he had confronted one of the most treacherous regions on the planet and walked away with a greater trust in himself and his path in life. While he did not have wealth, like Osborn, and had not yet discovered a major species, like Hatcher, he had tested himself in a grand arena and left its stage pleased with the results. That his finds had come after a man he looked up to as a mentor decided to give up only sweetened the triumph and made him eager to see how far he could go.

He gathered the self-assurance to write another letter to Osborn, this time approaching him as an equal and demanding honesty in return. His time in Patagonia had made clear what he knew all along: he was not meant to be an academic, with one foot in the classroom and one in the field; he was built for a life of action, as comfortable in the roughest terrain in the world as he was carousing in the bars of Greenwich Village. "I have been with the Museum now three years at $50 per month and assets have just about covered liabilities," he wrote. "Although I failed at Columbia University (it's a bitter pill to swallow) my aim there was not lost. I tried to cover too much and got swamped. For me to remain with the Museum I am sure I can best serve its interests in the field where physical energy and resource are most called for. After a thorough collection has been made from the different horizons in South America . . . There is South Africa, Australia and Siberia which must eventually be represented in the American Museum. But this takes time and means, if I am the man to do the work, that I must give up other projects and interests and rely wholly on my salary from the Museum," he wrote, adding that he "[did] not feel justified in doing this for less than $100 per month . . . believe me sincerely this letter is not dictated by a spirit of greed but by an awakening that I must know where I am at." Osborn's response does not survive, but apparently it was enough to convince Brown to remain with the American Museum, where he would work for the next forty years of his life.

His future set and his fossils packed and sent on their way, Brown returned to Punta Arenas only to discover that no ships were scheduled to leave for New York for another four months, extending his time in South America once again. Loath to take on any new expenses without Osborn's approval, Brown "decided to do some exploring," he wrote. Fun had a way of finding Brown, and he soon fell in with a quasi-professional gambler named Saltpere, who was never seen without gloves on and could recite several full novels by Charles Dickens from memory. In one of the more bizarre episodes of a life that was full of them, Brown begged Saltpere to take him on a joy ride around Cape Horn after the man casually mentioned that he owned a six-ton boat. Saltpere agreed, but told him they would first need to work together to recover the boat from a one-armed man who had stolen it and moored it near a shipwreck off Tierra del Fuego when he realized he couldn't sail it on his own. Brown couldn't tell how much truth was attached to Saltpere's tale, but decided to go along with it to see where it led. Sure enough, a few days later Brown stood next to Saltpere as he confronted the one-armed thief on the stolen boat and retook possession of his vessel. Having secured what was his, Saltpere offered to take the thief along with them on the trip to Cape Horn and was disheartened when he refused (a smart move, given that Saltpere was likely to leave him on an island as punishment, Brown later surmised).

Saltpere, Brown and a crew member continued south toward the Cape. While in the Le Maire Strait, the three men spent a turbulent night battling the wind and fearing that they would be taken far out to sea. On another night, when the water was calm, Brown lay awake on deck deep into darkness, marveling at an overpowering feeling of connection with the Earth. "What a sight was presented in the pitch-black night—schools of porpoises cutting back and forth across the bow, leaving a streak of silver in the phosphorescent medusa-filled sea," he wrote. His sense of foreboding around the ocean never quite left him, with good reason. Not long after, Brown was exploring the shoreline in a canoe when another sudden storm struck. He paddled back to the larger boat and grabbed the wheel, attempting to maneuver it through the narrow bay while Saltpere and the other crew member secured what they could. The boat crashed into a rock and ripped six feet off its side, tossing Brown overboard into water so cold he couldn't swim. As he felt his body seize up, he grabbed a barrel that had also been flung from the sinking boat and rode it to shore. Saltpere and the crew member survived by grabbing onto scattered shards of the doomed vessel, and they soon washed up on the same rocky beach as Brown. With no boat, no supplies and no way to send for help, the trio walked north along the remote coast for days until they came upon a camp of startled Australian gold miners. They waited there another sixteen days before they were able to signal to a passing ship, which took them back to Punta Arenas.

✣ ✣ ✣ ✣

At last, Brown was ready to leave Patagonia. He had survived a shipwreck, killed a mountain lion, wrestled an octopus and outlasted the mentor whom he considered one of the toughest and most talented prospectors he had ever known. The haul of fossils he sent back to New York confirmed his worth without question. Today, several specimens that Brown collected on his solitary expedition in Patagonia remain on display on the floor of the American Museum, including the skeleton of a primitive ground sloth known as *Hapalops* and the bones of three species of ungulates, a group of animals that were the forerunners of modern horses, camels and deer. In the wild with no one but himself, Brown faced down his self-doubt and now looked forward to rejoining his life in New York and the possibility of reestablishing a relationship with Marion, if she remembered him. If she did not, then the trip was still worth it. "For many months I had been out of touch with civilization," he wrote. "There are no cables, and mail often reached me via Liverpool. The Spanish [American] War had been fought and won, but I was happy following the life work I had chosen."

The next scheduled ship to New York was several weeks away. Hearing of his predicament, the captain of a ship heading to Portugal offered Brown a ride across the Atlantic if he was willing to sleep in the ship's saloon. There, perhaps, he could find a quicker connection to New York. Brown soon found himself in Paris and London, examining the collections of what were considered the best natural history museums in the world, the dirt from his digs in Patagonia still clinging to his boots. He spent his days inhaling culture in art galleries and at monuments, and his nights thirsting for adventure in bars and salons.

Brown finally returned to New York having shed his nasal Kansas accent and all other markers of the farm boy he once was. "Since he had lost all his clothing in the shipwreck near Spaniard Harbor, he bought an entire new outfit in Paris, including a tall, collapsible silk hat," his daughter would later write. "Sporting a moustache and a pointed Vandyke beard, he then considered himself the epitome of a cultured European."

For the rest of his life, Brown brought a suit, tie and full-length beaver fur coat on digs, often looking more the part of a man ready to drop into an opera than a scrappy paleontologist. It was as if he was forever making up for lost time, trying to will away those hours of suffocation and containment on the farm by becoming the flamboyant explorer of his boyhood dreams. That self-possession would soon meet its gravest challenge, however, as Brown realized upon his return how much New York—and indeed, his standing at the museum—had changed in his absence.

Chapter Nine

BIG THINGS

BROWN WATCHED THE TOWERS OF LOWER MANHATTAN
grow larger as the steamship carrying him back from Europe passed
through the Verrazano Narrows on June 10, 1900. He had been
gone for sixteen months, and in that time the promise of New
York had only grown. Its riches were larger; its population greater;
its possibilities wilder. It was the jewel of the world's fastest-grow-
ing economy, drawing the most talented and ambitious of a young
nation. Nowhere else was the link between science and wealth as
highly regarded, and nowhere else were the profits of an economy
built on industry and machinery displayed as lavishly. Skyscrapers
soared and mansions bloomed, each one a new ornament to prosperity. The American Museum was among them, an intellectual
cathedral carved from the city's wealth. Brown looked forward to
spending several months reacquainting himself with the museum's
collection and making space for the specimens which he had collected over his long absence. And in his personal life, he planned to
contact Marion, hoping that their relationship could rekindle after
a long period of silence.

Within weeks, however, he was in the badlands of South Dakota
and Wyoming.

✢ ✢ ✢ ✢

NEARLY AS SOON AS BROWN stepped off the boat, he found himself well behind in a race that he did not know had begun. When Brown last spoke with William Reed four years earlier, the veteran fossil collector was enjoying the novelty of steady employment with the University of Wyoming after his long and often prickly relationship with Marsh. Perhaps feeling generous to a young new collector, Reed had helped Brown locate the fossil beds where he uncovered the American Museum's *Diplodocus*, and asked for nothing in return. There seemed to be little need. Through his work, the University of Wyoming built a collection of dinosaur fossils that rivaled any on the East Coast, while Reed seemed to finally have contentment within his grasp after a lifetime of chasing dreams.

What the university did not have, however, was money. Through his work in remote and challenging terrain, Reed had helped create a new economy dealing in large bones. Forever a prospector at heart, he ultimately could not find it within himself to accept the security of a regular paycheck when there was a flicker of a large payday ahead of him. "Our university is so poor that I am thinking of leaving it and selling my fossils in Europe or to some other American museum," Reed confided in a letter to Marsh, hoping to spark his interest in a purchase. His funding dry and his reputation wounded after his behavior in the Bone Wars had fully come to light, Marsh was in no position to take Reed up on the offer. Reed, however, was not dissuaded. "The bones here are no. 1 in quality and some of them are monsters," he wrote. Like a man already planning in his head how he would spend his winnings from a lottery, he confided to Marsh that he was "writing to several large museums . . . in hopes to work up a market for this material before spring."

Invitations to tour his workshop and marvel at his discoveries soon reached newspaper reporters in every city with a major museum. The *New York Journal*, owned by newspaper baron William Randolph Hearst, devoted a full page to announcing that "Most Colossal Animal Ever on Earth Just Found Out West." The article appeared on December 11, 1898, just four days after Brown left the city bound for Patagonia. While Brown was battling seasickness, countless New Yorkers examined the newspaper's illustration of a streetcar attempting to stop itself from crashing into the tail of a *Brontosaurus* standing on its hind legs as it peers into the eleven-story windows of the stately New York Life Building on Broadway. "When [the dinosaur] ate it filled a stomach large enough to hold three elephants . . . when it was angry its terrible roar could be heard for ten miles . . . one man cannot lift its smallest bone," the paper reported. A portrait of Reed standing next to a femur taller and wider than his own body ran with the article, along with a drawing of what the beast's skeleton would look like when fully assembled. In it, the dinosaur towers over a man standing underneath its belly, the size difference suggesting a dog and one of its ticks.

A copy of the article landed on the desk on one of the wealthiest men who ever lived and who, as luck would have it, was then in the process of trying to fill up a natural history museum he had recently established in the town that had made him rich. "Can't you buy this for Pittsburgh?" Andrew Carnegie wrote in the margin of a copy of the article he forwarded to the director of the Carnegie Museum of Natural History. "Wyoming State University isn't rich—*get an offer*—hurry."

Andrew Carnegie had not always been interested in big things. The son of a handloom weaver and a shoemaker who emigrated from Dunfermline, Scotland, he dreamed as a boy of becoming a bookkeeper. With his uncle's help, he got a job running messages for the O'Reilly Telegraph Company. There, he found that he was comfortable in the often violent and ruthless world of industry and willed himself into a larger life. By the age of twenty-four, his mastery of accounting and the complexities of the railroad schedules helped him rise to a position as a district manager at the powerful Pennsylvania Railroad.

Stuck with a body whose height did not match his ambition, he began wearing high-heeled boots and a top hat to make himself seem more imposing than his natural five feet suggested. By the age of thirty, he was running his own company, with several of his former bosses at the railroad as secret partners, and grew rich from now-illegal insider contracts which paid him handsomely to supply the company with raw material. By forty, he controlled steel companies, iron ore mines, oil wells, bond trading firms and bridge builders. He spent his days in a hotel suite in Manhattan, where he worked only a few hours each morning yet reaped unimaginable wealth. "Ashamed to tell you profits these days," he wrote to a friend. "Prodigious!"

Carnegie was the rare captain of industry for whom each new blessing felt like a burden. Though seemingly every move he made in his business life proved in time to be the right one, he grew tired of the hunt for profit and tried to fashion himself into a man of culture, forging friendships with artists and intellectuals at home and in Europe. In his thirties, he resolved to "make no effort to increase my fortune, but spend the surplus each year for benevolent purposes." Philanthropy, he argued, was "the true antidote for the temporary unequal distribution of wealth, the reconciliation of the rich and the poor." Over time, he built 1,689 public libraries in the United States, set up a trust to pay the tuition of Scottish university students, funded pensions for American college professors, established a Hero Fund to award civilians who took extraordinary risks to save another person's life, built a complex of four world-class museums in Pittsburgh and founded a think tank to promote international peace. "The man who dies thus rich dies disgraced," he wrote in an essay called "The Gospel of Wealth," published in June 1889. A museum of natural history provided the unique prospect of increasing interest in science among those who had not had the opportunity or inclination toward formal education, fueling the sort of intellectual self-improvement that Carnegie saw as the key to future prosperity. In a city such as Pittsburgh, bursting with people who worked with their hands, a museum that did not "attract the manual toilers, and benefit them . . . will have failed its mission," he wrote.

Dinosaur bones, especially those so large as to make an adult tremble, were exactly what he needed to bring in the masses and kindle their curiosity about the wider world. Though the field of paleontology had fully separated itself from geology and was becoming increasingly specialized as it branched off into what would become the subdivisions of paleobotany and paleoecology, a wide gulf remained between what scientists understood about the prehistoric world and the public conception of the deep past. The few displays of dinosaur fossils or their lifelike models in the United States and Europe positioned the creatures largely as million-year-old novelties, spending little time on the reasons why these extinct creatures mattered to science or our understanding of the planet. Dinosaurs at the time were largely considered spectacles and nothing more, as if their size overshadowed any need to explain their significance. Yet with a display of impressive extinct creatures in a museum setting—the larger the better—Carnegie believed he could nudge others along the same path that he had once walked. A childhood interest in the dismal science of accounting had unlocked the traits which he rode to fantastic wealth; for the child of an ironworker, there was no telling where the spark ignited by an exposure to dinosaurs and all they represented could lead.

William Holland, the director of the Carnegie Museum, knew almost nothing about paleontology. The son of missionaries who served in Jamaica, Holland trained at the seminary and worked as a pastor in Pittsburgh before turning his attention to entomology. Like his counterpart Osborn at the American Museum, his greatest strength was unwavering self-confidence. Doubt that he was the right person to lead a campaign for dinosaur bones seemed never to enter his mind, and he faced the world as if he were simply waiting for validation of his greatness. He contacted Reed shortly after the article in the *New York Journal* appeared and was told the dinosaur was not for sale. Not long after, however, Reed followed up with a letter saying that the University of Wyoming owned the specimen depicted in the article, but that he personally owned another quarry containing a fossil just as large, if not larger. Sensing his payday had finally come, Reed described it as "a good prospect, the best I have seen for many years," yet stressed he was under a time constraint: his success at finding fossils had convinced the Wyoming state legislature to appropriate more funds to the university to expand its collection, and it would be difficult to turn the regents down if they offered a fair price. Unless Carnegie acted fast, he would lose his "chance to get this monster," Reed warned.

Holland wrote back at once, laying out exactly what he wanted. "I am anxious to know from you how perfect this skeleton is likely to prove," he wrote. "Do you think that it promises to be as perfect, for instance, as the skeleton of the Brontosaur which Prof. O. C. Marsh has at Yale? . . . I should like very much to obtain for this Museum a specimen as nearly perfect as possible of one of the huge saurians. . . . I should dislike however to enter the arena as a competitor for a mere fragment of these remains, no matter how large they may be, because I set our sights to erect one of the largest specimens in our Museum Hall." Reed assured Holland that the fossils would prove to be exhibition-worthy. Holland boarded a train to Wyoming and presented Reed with a lucrative offer, which included a three-year contract to join the Carnegie Museum staff. Only after Reed signed the paperwork, however, did he disclose the fact that the University of Wyoming had an ownership stake in the quarry where his prospective dinosaur lay and it was not his to sell outright.

Not wanting to tell the world's richest man that he had been bamboozled by a former shepherd with no formal education, Holland tried to buy his way out of the problem. He offered the university regents $2,000 for the fossil but was told that "it is the biggest thing on earth and we think that it is worth a hundred thousand dollars." The rejection further convinced Holland that such an immense prehistoric relic belonged in a major metropolitan museum where the greatest number of people could see it, and not in the university collection of a lightly-populated state. Holland hired attorneys and bought up land and mineral rights, looking for a way to block the university's claim to the material. Though he too wanted the dinosaur, Carnegie feared that Holland's aggressiveness would turn into a public relations disaster that he could not afford. A battle between striking workers and hired private detectives at his Homestead, Pennsylvania, steel plant eight years earlier that left seven dead had severely damaged his reputation as a friend of the working class, and his focus on philanthropy was one attempt to bandage that wound. Holland turned to lobbying the state's governor for assistance and made a second trip to Wyoming to pay local cowboys for any leads about other fossil outcrops. "We shall ultimately get possession of our coveted monster," Holland wrote.

On a trip to New York to confer with Carnegie, Holland met Jacob Wortman, then on staff at the American Museum, and spoke with him about the difficulties of fieldwork in Wyoming. Wortman made such a strong impression on Holland that he offered him a job as a curator at the Carnegie Museum, where he would be free of the overbearing shadow cast by Osborn. Shortly after Wortman accepted, he traveled to Wyoming to check on the quarry holding Reed's find. There he discovered that amateur workmen hired by the university had removed most of the fossils that Reed had identified and destroyed the rest. Holland tore into the university regents, who, he claimed, "do not know how to meet manly men in a manly way, but are as full of little narrow, petty jealousies as an egg is of meat." Empty-handed, he convinced Carnegie to fund a full team of prospectors to Wyoming in the coming months, where he planned to have Reed find a fossil that would justify all the headaches he had already given him. "It is . . . of the utmost importance that our Museum should succeed in obtaining a fine display of showy things," Holland told his field crew. "Mr. Carnegie has his heart set on Dinosaurs—'big things'—as he puts it."

THE CARNEGIE EXPEDITION WAS NOT the only one descending on Wyoming in the summer of 1899. Aiming to open up travel on one of its least-utilized routes, the Union Pacific Railroad offered free passage to the state for a limited time to amateur prospecting teams from over two hundred colleges and universities. Known as the Wyoming Fossil Fields Expedition, the invitation promised "the fields are ample for all who wish to avail themselves of the opportunity to collect and create museums as large, if not larger than any that have been built up during the last quarter of a century." For a brief instant, paleontology was no longer restricted to those who had the ability to foot the bill for travel to the far reaches of the country. The inherent advantage enjoyed by big city museums like the American Museum and the Carnegie would for the first time be tested on an open playing field, where skill and luck at uncovering fossils would matter more than a reservoir of money. "I cannot help thinking that this miscellaneous scrambling for dinosaurs . . . will be a misfortune," Osborn noted, no doubt uneasy at having to participate in a competition in which his status did not give him an unfair advantage.

Soon, the wilderness was dotted with those who dreamed of finding dinosaur bones. Some were backed by money; a team from the Field Museum in Chicago could draw from a $1 million endowment donated by the department store magnate Marshall Field. Others were college students hoping that a find could secure a foothold for their career, or at the very least give them bragging rights when they brought something impressive back to campus. And there were still others—cowboys, drifters, down-on-their-luck gold miners—who only vaguely knew what a dinosaur was or how you would go about finding one, but who recognized that anything that brought so many educated people out to Wyoming must be worth something. They, too, joined in the chance to strike it rich.

In one summer in one of the most remote parts of the country, the dawning age of science, industry and professionalization faced off directly against the era of hard living, grit and luck it was replacing. "The old-time expeditions were staged in the real West, at a time when lack of means of transport . . . together with the very intimate contact every fossil hunter must have with his physical surroundings—with fatigue, heat and cold, hunger and thirst—made the research for the prehistoric a real adventure suited to red-blooded men," a professor of zoology at the Massachusetts Agricultural College who prospected in Wyoming that summer later recalled.

Prospectors rich and poor descended on Medicine Bow, a desolate town on the plains whose train station, three saloons and twenty houses were the only permanent structures in the sweeping valley of sagebrush and stone. "Until our language stretches itself and takes in a new word of closer fit, town will have to do for the name of such a place as was Medicine Bow," wrote Owen Wister, a Harvard graduate and close friend of Theodore Roosevelt whose novel *The Virginian*, published three years later, was set in what locals called the Bow, and is now considered the first Western. "I have seen and slept in many like it since. Scattered wide, they littered the frontier from the Columbia to the Rio Grande, from the Missouri to the Sierras. They lay stark, dotted over a planet of treeless dust, like soiled packs of cards. . . . Yet serene above their foulness swam a pure and quiet light, such as the East never sees; they might be bathing in the air of creation's first morning."

Medicine Bow was a dusty and dirty remnant from a vanishing time. Ten years earlier, the U.S. Census Bureau had announced that the frontier was closed, ending the nation's conception of itself as an ever-expanding force across the continent. Yet there still remained a large gap between what government officials considered settlement and what life was like on the forgotten plain. "We have no law here," warned Holland's landlady when he arrived to lead the Carnegie team. A member of the American Museum staff later confessed, "Medicine . . . is a little bit the worst town I have ever seen. . . . I would advise no one to spend a night here if he can help it." It was a place where lawlessness could literally land in your lap. Butch Cassidy and the Wild Bunch, a gang of outlaws and bank robbers famous for their brazenness, held up a train and dynamited its safe, containing what would now be worth more than $1 million, in the mountains outside of Medicine Bow shortly before the paleontologists—professional and not—reached town. Once the gang was safely gone, locals picked up the scraps for souvenirs. Holland was given one as a present his first night in town, and he promised to add it to the Carnegie collection.

The few rooms in Medicine Bow fit for human habitation quickly filled up with workers straightening and regrading the Union Pacific tracks spanning the prairie. Yet still more people came. Trains carrying more than one hundred members of the Fossil Fields Expedition arrived over the span of a few days. While most groups were small, those who could afford them brought along teamsters and cooks and field hands, swelling Medicine Bow's population well past its capacity. Soon, nearly every flat place to sleep under a roof was claimed. A run on the town's general store emptied it of every pickaxe, bag of flour, hammer and chisel. A professor from the University of Kansas who arrived a few days behind the crowd was shown to a room filled with mice crawling over a bloody mattress, prompting him to bed down in a railroad equipment shed instead.

Fossil hunters braved the dangers of Medicine Bow because of its proximity to the Freezeout Mountains, visible at the edge of the horizon. Smooth, gray and nearly treeless, the range is made up of sandstone and shale and conceals abundant beds of Jurassic fossils in a layer of rock referred to as the Morrison Formation. Named after the town of Morrison, Colorado, where it was first discovered, the Morrison Formation dates to about 150 million years ago, when what are now considered the badlands of Wyoming, Colorado and Utah were crisscrossed with rivers, streams and ponds.

In what was one part science and one part gold rush, teams of prospectors—some carrying the flags of their colleges or institutions, nearly all of them armed with rifles and other weapons—swarmed over the mountains and attempted to claim as many quarry sites as possible without knowing what they held. Those who should have never been in the mountains in the first place soon gave up, overheated and exhausted with nothing to show for it. Others tumbled off rocks or tripped over tree branches, their grand adventure cut short by a broken nose or busted ankle. Few knew how to read the rocks of the Morrison Formation. Success was largely a matter of matching small variations in color with knowledge of what the rock had once been. Dark gray indicated volcanic lava, which could be safely ignored. Grayish-green rock represented layers of siltstone, which likely had been the bottom of a river or a sandbar—exactly the sort of place where the body of an animal could wash up and become covered in sediment.

Even those who had a cursory ability to fend for themselves in the wild often came out empty-handed, unfamiliar with the curves of the mountains. Prospectors pushed deeper into the inhospitable terrain, hoping to shake themselves of the dilettantes and the desperate. The days "consisted in 'looking out' as much as one could cover of the rock formations exposed in canyons and gullies, prowling over the weathered slopes, and climbing along the steeper cliffs, watching always for the peculiar colors and forms of weathered bone fragments, following up every trail of fragments to its source, and prospecting cautiously with a light pick or digging chisel to see what, if anything, is left in the rock," wrote William Diller Matthew, a curator at the American Museum who spent three weeks that summer in Wyoming. "Such a prospect came often as a blessed relief after hours of climbing and scrambling had reduced one to a state of staggering weariness. . . . Then supper, a pipe, and to bed, and the same routine repeated the next day and the next."

The Fourth of July provided one of the few breaks in the monotony of the search, a clear reason for this particular day to be different from the rest. Throughout the big, open states that made up what was becoming known as dinosaur country, the holiday was considered a momentous event, one of the few outward signs that there could be any connection between this rough land and the crammed cities on the coast. Prospector camps were typically full of booze and hard living to begin with; the Fourth of July took it to another level, a chance to prove that the prospectors had shown restraint on every other night of drunken carousing.

While the other members of the Carnegie team downed liquor and beer, set off fireworks and gambled in camp, Reed set off alone to hunt, the noise of the camp serving as the breadcrumbs that would keep him from getting lost. He found antelope footprints and began tracking the animal. It was then that he noticed a bone sticking out of a rock. He began digging with the small tools he carried with him and realized that he had found something.

As the Carnegie team nursed their hangovers the following morning, Reed revealed that he had stumbled upon a section of what appeared to be a *Diplodocus*. The others joined him at the site, and it soon became clear that they were at the start of a major find. Onlookers from competing museums began to appear each day at the excavation site, drawn to the spectacle of watching an 84-foot-long prehistoric beast emerge chisel by chisel from the matrix of rock. Over the following weeks, the specimen revealed itself to be one of the most complete *Diplodocus* ever found, containing all major bones and a flawless skull. The team sent word to Holland, who seemed not to recognize how rare it was to find a nearly complete specimen and pressured Reed not to come back until he had secured the animal's full set of bones. "My dear fellow, the rest is all there, and do not you forget it," he wrote.

No amount of second-guessing could take away from the importance of the find. Though he was not personally responsible for uncovering it, Wortman was particularly pleased that the Carnegie, and not the American Museum, had claimed the most impressive discovery that summer. Not only that, but his team had recovered other fossils that would fill out the Carnegie's halls faster than Holland had anticipated. "We have labeled our quarries and I do not apprehend any difficulty in holding them until we can get around to work them out," Wortman wrote in a letter to Holland. "There are many parties here already and they are getting pretty short on bones. The quarry of the American Museum is turning out badly and they are looking for other locations. Me thinking [sic] they will get little. They lack experience."

When news of the Carnegie's find reached New York, Osborn arranged one of his rare trips to the field, unable to rest in the knowledge that his men were falling behind. He reached Medicine Bow a few weeks later and rode out on horseback to the quarry site. There, he dressed down Walter Granger, his well-liked and respected field foreman, and second-guessed every decision he had made. Among his worst sins in Osborn's eyes was his decision to focus on a few sites rather than widen his approach and outwork his rivals to the point of exhaustion. "I reached here yesterday evening after an interesting ride across the plains and my arrival is on time for I find matters somewhat disorganized," Osborn wrote in a letter to his wife. "We have just taken out a very fine specimen, but have no more in sight, while Dr. Wortman after two months of very bad luck has made a ten strike, finding an unusually fine skeleton, just what we needed, in fact. I shall put revived life into the party and thoroughly reorganize their work." The element of luck in Reed's discovery was not lost on Osborn, yet he could not accept that a rival was pulling ahead. The realities of camp life were too rough for him, however, and he left after a week to vacation in Colorado, where he continued to send letters to Granger with new commands.

The summer of 1899 proved to be one of the most influential concentrated efforts to dig out dinosaur bones in the short history of paleontology. The amateur collectors who braved the field thanks to the free passage given by the Fossil Fields Expedition unearthed more than two tons of fossils, scattering specimens to universities and colleges across the country, where they offered many people their first opportunity to see clear evidence of prehistoric life. The Field Museum, meanwhile, brought home to Chicago seventy-five dinosaur bones weighing a collective five tons, including the spine and pelvis of a *Brontosaurus*, twenty-five tail vertebrae from a *Diplodocus*, and the pelvis and foot of a *Creosaurus*, a predator now known as *Allosaurus* that grew up to 35 feet long and may have used its head as a hatchet, slamming it into prey and then ripping off flesh into a mouth filled with backward-curved teeth that prevented anything from escaping. For its efforts in the field, the American Museum amassed 131 bones for the collection, including half of a *Brontosaurus* skeleton. Above them all stood the Carnegie, which left Wyoming with most of the spine, pelvis, skull and eighteen ribs of the prized *Diplodocus*, as well as fossils of marine reptiles that expanded the breadth of its shelves. "We obtained a quantity of material which would have made the mouths of Cope and Marsh water," Holland wrote in a letter to a friend.

For the first time since he pivoted his life to focus on paleontology, Osborn had come in second on a public stage. Though the American Museum had made several important finds, nothing in its haul could compare to the specimens obtained by the Carnegie. "We have had a successful season's work, not brilliant this far," Osborn confessed in a letter to a friend. What's more, the *Diplodocus* the Carnegie brought home was both bigger and better preserved than the American Museum's specimen that Brown had discovered two years earlier, leaving what had been one of the most important specimens in the museum's collection diminished in Osborn's eyes.

The relationship between Holland and Wortman deteriorated over the winter of 1899–1900, inflating Osborn's hope that the threat of the Carnegie Museum would fizzle away. But things only got worse for the American Museum once Wortman left its rival. In Wortman's place, Holland hired John Bell Hatcher from Princeton, putting an even more experienced fossil hunter in charge of expanding the Carneige's collection of dinosaur bones. In his first months in his new job, Hatcher realized that the *Diplodocus* specimen uncovered in the summer of 1899 was not just immense, but represented a new species. He named it *Diplodocus carnegii* in honor of the museum's founder. In a move that made the discovery sting all the more, Holland asked Osborn as a matter of professional courtesy to confirm Hatcher's interpretation of the find. Holland knew that it was correct, yet could not resist forcing Osborn to publicly acknowledge another museum's superiority. "When it comes to scientific proposition, you will permit me to say that I will bet on you every time against Osborn," Holland wrote in a letter to Hatcher. "However, this is of course said confidentially."

Holland wasted no time in declaring that the Carnegie Museum would soon put Pittsburgh on the world stage. "To the fame of Pittsburgh as the seat of some of the most Cyclopean industries of the age is being added reputation as a seat of learning," he wrote. "Under the cloud of smoke, which attests the industry of her inhabitants, and is the sign of her material prosperity, live men who find their pleasure in exploring the wonders of the material universe, and the record of their discoveries and researches will from year to year be found in the Annals and Memoirs of the great Museum which the more than princely generosity of Mr. Andrew Carneige has called into being."

AS BROWN WALKED INTO the American Museum after eighteen months away, he knew little of the threat that the Carnegie Museum posed to him or his institution. He had left New York as the hero who found the museum's first dinosaur fossil; he returned to learn that his discovery had been usurped by a larger, better-preserved specimen of a newly-recognized species which the man who left him behind in Patagonia had named in honor of the one person who could outspend the American Museum to obtain fossils. It was as if all of Brown's fears had come true at once. He was missing when it mattered the most, and now his work uncovering mammalian fossils in Patagonia seemed inconsequential, the jaunt of an unfocused boy rather than the stern focus of a professional like Hatcher. His sense that time was passing him by was heightened after he visited his family on the farm in Carbondale on the way to the prospecting fields that summer and found his father "very feeble," as he later wrote.

Should he not return with a comparable fossil that summer, Brown feared that Osborn would use his ample funding to find someone who could. He would not be the first person discarded by Osborn once he had served a purpose, nor the last. From his perch in New York, Osborn continually rode his field prospectors to be more aggressive, casting wide nets to lay claim to many quarries rather than honing in on any one site until it was clear it contained a major find. "So I say once more prospect, prospect, prospect, prospect," Osborn wrote in a letter to Brown years later. "Granger was digging instead of prospecting when he let Wortman slip in and find that big *Diplodocus*."

Brown set off to the badlands, hoping to find a *Triceratops* skull that would be fit for exhibition. He landed in the broken prairie of South Dakota, about forty miles west of the small city of Edgemont. There, he tracked streams branching off the Cheyenne River, following the path of a floodplain that existed 66 million years ago through modern-day cattle ranches. Over the following weeks, he found several broken fragments of *Triceratops* skulls, building a small collection of their horns, yet nothing that could be useful for the museum.

In Patagonia, he had had the freedom to explore where his curiosity took him. Now, he felt the presence of other collectors, even when they were hundreds of miles away. That summer, Hatcher was in Wyoming, where he was focused on not only unearthing the remainder of the immense *Diplodocus* first identified the year before, but was intent on drawing on Carnegie's limitless funds to find further prospects before rivals from the American Museum or Field Museum could. "I do not care to have you . . . say anything to your [scientific friends] about our plans at present. . . . We do not care to be trailed, and we are keeping our plans quiet," Hatcher wrote in a frank letter to Carnegie before he headed out to the field that summer. In Colorado, meanwhile, Elmer Riggs, a curator at the Field Museum working on a dig near Grand Junction, uncovered a thigh bone on July 26, 1900, which measured six feet ten inches in length, "longer by eight inches than any limb-bone, recent or fossil, known to the scientific world," he wrote.

After word of Riggs's discovery leaked out, newspapers across the country ran articles about what the *Boston Journal* called "The

Monster of All Ages." Riggs's work on uncovering the rest of the beast—now known as a *Brachiosaurus*, a giraffe-like sauropod that was one of the few dinosaurs whose forelimbs were longer than its hind legs—became so much of a local attraction that he begged the townspeople to keep it to themselves. "There are half a dozen parties collecting fossils in the west who would eagerly turn to such a new region if they had wind of it. They must not have opportunity to learn, through the press or otherwise, until the end of the season when we shall have had time to prospect the valley and take the cream of it," he wrote. Yet the tourists kept on coming, drawn by the slow process of unburying a giant whose existence defied the imagination. "I enjoyed having visitors and took pains to explain things to everyone interested, especially those of intelligence, but when the idlers began coming just to have some place to go, and souvenir fiends began prying off pecans whenever we were not around, it became tiresome," Riggs wrote. "The souvenir hunters mean no harm to be sure, but as one man put it . . . they would 'steal the halo off of Christ's crown!'"

Brown, by comparison, worked alone and in obscurity. He continued his hunt throughout the long summer, covering hundreds of miles on horseback without any significant bones to show for it. Even in the wilderness, he learned of his competitors' successes in dispatches from New York that reminded him of the stakes he was facing. "Report from Pittsburgh is that Hatcher has had a wonderful season's work, sending back three car loads of fossils, if this is true, he has probably done better than we have," Osborn wrote. Each day of failure compounded on the last until soon Brown questioned why he was there and whether his long luck had finally run out.

Then, on the first day of September, he uncovered a nearly complete duckbilled specimen the size of a rhinoceros. "It gives me great pleasure to announce to you the discovery of a *Claosaurus*," Brown wrote to Osborn, no doubt hoping that his enthusiasm would paper over the fact that it was not as large or as impressive as the discoveries that summer by Hatcher or Riggs. Marsh had discovered the first known *Claosaurus* specimen in Kansas in 1872, diminishing Brown's find in comparison to the newly-unearthed monsters in Wyoming. Still, it took him a full month to extract it from the sandstone. The fact that it proved to be an exhibition-worthy specimen did little to burst Brown's ballooning sense of disappointment. He continued to roam the prairie, looking for the elusive *Triceratops* skull that he knew Osborn desired. Such a find would not only reinforce his standing at the American Museum after his long absence, but would maintain the reputation and relevance of the institution at a time when it appeared to be falling behind in the face of competition from titans who saw dinosaur fossils as a chance to prove their superiority to their counterparts in New York. Failure, meanwhile, would push Brown one step closer to losing all that he had worked for. He had claimed a foothold in New York, but he knew how precarious that position was.

He remained in the field throughout the fall, undeterred by the dropping temperatures. A sudden blizzard interrupted him one day in November, yet he refused to let the seasons hold sway. Osborn, normally content to benefit from the labor of others, told Brown that it was time to return to New York. "I have been fearing the snow storms would overtake you but I congratulate you upon the continued success of your work. . . . I think you had better pack up and come in as soon as you can."

Still, Brown soldiered on. The obstacles only mounted. One morning, he woke up to discover that his team of horses had escaped from camp to look for warmth, forcing him to search for five days on foot before he finally found them fifty miles away. Later, when he began hauling some of his finds out from the field, his wooden wagon broke under the strain of two tons of fossils, collapsing the load but fortunately not breaking any of the bones in the process. "I got out of the wreck with only a few bruises,"

Brown later wrote. Thanksgiving passed, and then Christmas, and then New Year's, and yet Brown searched on.

A deep freeze in early January finally convinced Brown to return to New York. As the wind whipped around him, he loaded thirty boxes of fossils into a train car of the Baltimore and Ohio Railroad. When some of the pieces of rock proved too large to fit inside, he helped railroad workers remove parts of the railcar's door to make room. As the train trundled toward New York, Brown could not help but console himself with the thought that his luck would return in the spring and bring him to the elusive *Triceratops* skull. While he had gathered several important specimens over the extended prospecting season that had just passed, he had no trophies he could point to that would remind himself—and Osborn—of his value. He had let himself down in the field for the first time in his life, and he hoped that it would be the last.

Chapter Ten

A Very Costly Season

For the first time in his life, Barnum Brown did not know where he belonged. As a child, the answer had always been simple: anywhere but on the farm. He spent his time at college obsessed with rising above the life expected of him, willing himself to find fossils that would open a path beyond Kansas and into the wider world. Now, having secured a toehold on his dream only to fail to keep up with a rival at a crucial moment, some of the first pangs of doubt began to creep into his conception of himself, taking over the space once occupied by his desire to be more than he was.

He had every reason to feel cast aside. At the start of the 1901 field season, Brown looked forward to returning to South Dakota, where he planned to redeem himself through the discovery of a *Triceratops* skull. Osborn, however, ordered him to Flagstaff, Arizona, to assist an expedition headed by the Smithsonian Institution that was searching for fossilized pine cones near the Grand Canyon ("great quantities of fossil wood" were available, Brown later wrote in a letter to Osborn, not quite concealing his lack of excitement at searching for something plentiful while far from the action). For dinosaurs, Osborn turned instead to George Reber Wieland, who had recently completed his doctorate at Yale. Wieland had once caught Marsh's attention after he uncovered a complete skeleton of *Archelon ischyros*, which at fifteen feet long was the largest turtle ever known to exist. He worked at the Peabody alongside Marsh until the professor's death in 1899, and now seemed in the best position to take over his mantle.

A short man with a booming voice and an infectious sense of purpose, Wieland easily wore the qualities Brown lacked: an Ivy League degree, a comfort with his status that allowed him to interact with Osborn on an equal level, and an ability to convince himself and everyone around him of his worth even when evidence was lacking. Osborn was so taken with Wieland that he readily accepted his steep demand for a salary of $150 per month, a rate nearly double what he paid more experienced collectors such as Granger. "I regret to tell you how important the financial side is for me," Wieland wrote to Osborn before agreeing to work for the American Museum that summer. "I wish you could feel disposed to name some better sum as wages, for I expect direct results." To justify his rate, he hinted that he had a lead on a *Barosaurus*, a plant-eating sauropod that can be distinguished from the more common *Diplodocus* by its longer neck and shorter tail, though no one has ever found its skull. Osborn placed Wieland in charge of the museum's dinosaur expeditions that summer, demoting field collectors who had been with the American Museum for several summers. "[Wieland] is very confident of success . . . you will therefore refer matters to him," Osborn wrote to Granger, adding that Wieland "seems to be a thoroughly nice fellow and I do not anticipate any friction." The new leadership in the field reflected Osborn's unflinching demand that his prospectors find specimens that surpassed those collected by the Carnegie and Field museums, restoring the American Museum—and by extension, his own reputation—to its rightful place atop its competitors.

Wieland took charge of a party that reached the village of Hulett in northeastern Wyoming on May 15. There, he established a makeshift camp on the west side of Devils Tower, a nearly vertical shaft of igneous rock that rises 867 feet above the surrounding forest of ponderosa pines and the banks of the nearby Belle Fourche River. The picturesque location was once the seabed of a shallow inland sea that began retreating 195 million years ago, leaving behind bands of dark red sandstone, maroon siltstone and gray-green shale that in certain areas date to the Jurassic period, some 135 million years ago. Approximately 50 million years ago, an immense shaft of magma—whether coming from a sudden explosion of a volcano or the slow result of millions of years of erosion—pushed up through the rock layers, leaving Devils Tower behind. Though it is built of hard rock, the shaft of the Tower is marked by long columns of what look like folds, leaving the impression that it is a giant drip castle made by a child on a sandy beach. Sacred to Plains Indians tribes, the formation was given its current name when Colonel Richard Dodge, a U.S. Army commander in charge of an 1875 military expedition, decided to translate what "the Indians call . . . 'Bad God's Tower,'" he wrote.

Wieland left New York having convinced Osborn that an impressive dinosaur specimen would be in his hands by the end of the summer. Yet over his first days in Wyoming, his bluster ran into the reality of the wilderness. The party found nothing while surveying the immediate vicinity of Devils Tower, undercutting Wieland's theory that the shaft of rock would contain fossilized sediment from several different ages and leave plentiful fossils exposed. He expanded the search more than twenty miles downriver, without meeting any success. The grind of digging and surveying and climbing over dangerous terrain only to come up empty-handed began to wear on the team. Each day that ended the same way let a little more air out of the bubble of inevitability that Wieland's boasts had built around himself. One Sunday in camp, a field collector killed a rattlesnake with his boot heel and saved its skin, making it the first and only specimen the expedition collected.

After two weeks, Wieland turned the party around and built an elaborate camp near the base of Devils Tower in preparation for a visit by Osborn, who was touring all the museum's fossil digs that summer from the luxurious comfort of a private Pullman car complete with a personal attendant and chef, which was a gift from the president of the Denver & Rio Grande Railroad. He had to leave them all behind to reach Hulett, however, which at the turn of the twentieth century remained too remote for modern travel. In Deadwood, South Dakota, Osborn boarded a train he called "as primitive . . . as you ever saw" to head deeper into Wyoming. At the end of the line, he boarded a boxy stagecoach that jostled and bumped as it took him the remaining thirty miles over the Black Hills. Osborn rarely braved such conditions. The fact that he was willing to travel to Devils Tower without the assurance that a valuable dinosaur had been located underscored not only his trust in Wieland, but his increasing desperation. "I am greatly interested in this camp for upon its success depends a large measure of our record for the season," Osborn wrote in a letter to his wife.

Wieland was not in camp when he arrived, the sort of unintentional slight that Osborn did not easily forget. Once Wieland returned, he hurried to take Osborn on a tour of what he had identified as potential quarry sites, trying to paint with words a more successful picture than the lack of fossils revealed. Each spot was overgrown with trees and grass, showing no evidence that Wieland had blasted or begun stripping off the top layers of soil and rock. Osborn, who had never conducted a dig on his own yet had toured enough of them to know what he should be looking at, grew enraged at the lack of progress. In full view of the members of the expedition, Osborn warned Wieland that he was not only wasting his valuable time, but putting the future of the American Museum in doubt with a wasteful expedition that had nothing to show for its great expense. Stung by the criticism, Wieland offered to resign on the spot. Osborn refused, unwilling to accept failure. He spent another night in camp, hoping for a miracle, before deciding that he had had enough. He suffered another five-hour stagecoach ride through a driving rainstorm to the nearest train station. Once there, he canceled the remaining stops on his tour of the West and headed immediately back to the East Coast, where fewer bad surprises lurked.

Sitting in his private Pullman car, Osborn searched for a way to fill a hole that seemed to get deeper by the day. The Carnegie Museum had but one multimillionaire to please; he had all of New York's aristocracy. That year, John D. Rockefeller, whose Standard Oil monopolized the oil business in the United States and made him the wealthiest person in the nation's history, had been elected a patron of the museum. Alongside him were the most influential and powerful men in the country, including railroad magnate George Foster Peabody, Archer Huntington—the adopted son of Collis Huntington, who had helped build the first transcontinental railroad and loomed as one of the most important figures in California—and Jacob Schiff, one of the most influential bankers of his era.

All wanted something to show for their money, a prize that proved the institution they had chosen to support was truly the best in the world. That spring, the museum's Department of Mineralogy opened a new exhibit made possible by Osborn's uncle J. P. Morgan, which included a stunning display of diamonds, emeralds and sapphires. Tiffany & Co. had originally collected and prepared the stones for the Paris World's Fair, helping to cement its reputation as the premier jeweler in America, if not the world. Once the fair was over, Morgan purchased the lot for the museum, a transaction he found far more palatable than funding another summer of dinosaur hunts that might turn up nothing. While few things could elicit the same awestruck response from the public as dinosaurs, the difficulty of finding huge bones opened the door to the question whether it was reasonable to assume that there were other specimens out there waiting to be discovered. Morgan, for all his millions of dollars in donations, had yet to see a complete dinosaur mount displayed on the museum's floor. Another year or two without results threatened to plant doubt in the minds of donors that dinosaur digs were a fool's errand, a search for something too strange and obscure to be worthy of serious attention.

With each day that passed, Osborn felt his place slipping further in the hierarchy of the museum. That year's annual report did not mention his Department of Vertebrate Paleontology until the twenty-sixth page, putting it behind updates on the museum's search for signs of ancient human life in New Jersey and news of a gift of more than one thousand butterflies to add to its permanent collection. When the report finally did turn to his department, it seemed to further broadcast his anxiety. "Professor Osborn not only contributed largely to the maintenance of field expeditions, as shown in the Treasurer's Report, but also spent his entire salary in promoting the work of his department," the report noted, revealing that Osborn had once again resorted to tapping his family's money in order to bolster his position. What was worse, there was little promise that things would turn around soon. While other museums were readying their displays of giant dinosaur bones, Osborn could only announce that his department was finishing a painted mural of ancient horses to display in the year ahead.

The train rattled east, bringing Osborn back to face the judgment of the only people he felt mattered. It would still be at least a year or two before the *Diplodocus* fossil discovered by Brown would be mounted for exhibit, and he had nothing to fill the gap. Over the long hours alone, he went over his options. It was clear that Wieland was not the answer he had been looking for. (Indeed, that summer marked one of the last times that Wieland would venture into the field to search for dinosaur fossils; after his falling-out with Osborn, he eventually returned to Yale and established a long career in the lower-stakes world of paleobotany, focusing his research on the palmlike plants known as cycads.) In his place, Osborn turned to William Reed, who remained a one-man marketplace for fossils after his deteriorating relationship with Holland prompted him to quit the Carnegie. "Still I am out for bones," the aging prospector wrote in a letter to Osborn that summer. "I have had good luck so far and found two things that I think you may want . . . everything I get is for sale." Osborn decided to take him up on the offer, and soon wrote to Wieland to inform him that his ill-fated Wyoming expedition was over. "I fear the Black Hills is a failure. . . . We have had poor luck and must make the best of a bad situation," he wrote.

He ordered Granger to leave Devils Tower and ride three hundred miles south to Como Bluff, where he was to inspect Reed's quarry and report back on its condition and possible worth. A few weeks later, Granger wrote with word that the site contained "first class" bones of an *Allosaurus* and a *Stegosaurus* that were still embedded in the rock, along with a few outcroppings of indeterminate fossils that suggested the presence of others that would be revealed with more effort. "There might be some excellent things here," he wrote—a clear break from his weeks alongside Wieland, which had yielded nothing more exciting than a skull of a *Brontosaurus* that proved unsuitable for exhibition because of damage caused by gophers burrowing beneath it.

Osborn soon approved four hundred dollars to buy the quarry, though it did not salve his wounded ego. Nothing the American Museum had found compared with the *Diplodocus* now in the possession of the Carnegie Museum. Even its scientific name, *Diplodocus carnegii*, reminded Osborn of his failure. He wanted nothing more than to bring Hatcher over to the American Museum and have him repeat his past successes. Perhaps then Osborn would be in a position to name a new species after his uncle J. P. Morgan, presenting his benefactor with an immortal gift that no material wealth could match. He yearned to blaze a trail in science, to bring honor to the family name. Instead, he was left retreading the work of others, dependent on a mercenary for fossils that summer. But, after paying Wieland's exorbitant fee, Osborn was not in a position to dangle enough money in front of Hatcher to pry him away from the Carnegie.

His hope was that he could wait and allow Hatcher's caustic personality to make his tenure at the Carnegie Museum a short one. Over the winter and spring, the veteran collector spent his first prolonged time in Holland's company as he familiarized himself with the Carnegie's collection and its exhibition plans. The two men—both single-minded in their pursuits, both convinced that they were intellectually superior to the other—clashed almost immediately, in person and in public. Hatcher took to writing scientific articles that focused on small errors in Holland's latest works, and was not shy in letting him know of his diminished opinion. "I am also much affected by the further abuse you saw fit to administer on Nov. 28, when you called me a jack-ass and a d——d fool," Hatcher wrote to Holland in a monthly progress report. "Such language, it seems to me, cannot but tend to destroy the harmony, enthusiasm & interest so essential to the welfare of the institution."

Though his own personality was likely to inflame Hatcher even more, Osborn believed he could convince the collector to trade one millionaire for another and come to the American Museum the following year. That would end all of these false starts, and finally bring in the gargantuan specimens that Osborn had grown to view as evidence of his own self-worth. He made a point of meeting with Hatcher in late summer, and came away hopeful that his problems would be solved. "I . . . look forward with great relief to next year when Hatcher can do the working and the planning for these expeditions," Osborn wrote in a letter to his wife. "It will be a grand thing getting him in the [American] Museum—he is anxious to come and I think we will get on smoothly."

Until then, he had to make do. To calm his sense that he was falling further behind, he sent Brown on small scouting missions late in the summer to investigate some of the dozens of unsolicited tips that the museum received each year. A high school principal named Willis T. Lee, who met Osborn in Wyoming during the Fossil Fields Expedition, wrote to tell him that he knew of a Jurassic outcrop on the eastern side of the Rocky Mountains, not far from his hometown of Trinidad, Colorado. Brown spent four days searching along the Animas River and came away with scattered bones but nothing significant enough to justify a full expedition. Though happy to at least be back in the dinosaur hunt after half a season searching for prehistoric wood, Brown yearned for a chance at finding the giant fossils that would redeem him. The *Brachiosaurus* that the Field Museum was still in the process of excavating was in Colorado, increasing the chances that there would be other impressive finds in nearby fossil beds. Brown longed to follow his instinct and explore rather than hunt down another dead-end report from a rancher who had likely confused a dinosaur fossil with the remains of a steer, yet felt he could only hint at it. "I don't know Riggs's exact locality," Brown wrote to Osborn. "It must be rich from the newspaper stories. I wish we might have a show at it."

Neither Brown nor Osborn knew at the time that the quarry, and indeed Riggs himself, would soon be available. The Field Museum was in the process of cutting its funding for dinosaur collecting in favor of geology, the preferred field of the museum's director. Riggs's dinosaur, which was at the time the largest specimen ever found, would stand as the last major dinosaur addition to its collection until nearly the end of the twentieth century. Paleontologists working at the museum protested the cuts to no avail, unable to convince their administrators that the coming era of mounted dinosaur displays would forever change how the public interacted with science and the popularity of natural history museums as a whole. "The authorities of the . . . Museum do not seem to appreciate the fact, for fact it is, that paleontology is becoming the chief interest in nearly all the great museums of the world - its absence will one day be greatly regretted," wrote Williston, Brown's one-time professor, who was collecting for the Field museum that season.

Osborn did not bite at Brown's unspoken request to hunt for a trophy specimen of his own. Instead, he ordered Brown to remain in eastern Colorado and collect the fossils of prehistoric mammals. Brown was not in a position to argue, so he directed his ambition back into his work. He soon found the skull of a three-toed precursor of modern horses known as a *Protohippus*, as well as impressive full skeletons of a prehistoric antelope and an extinct horse known as *Hypohippus*, which was distinguished by its long neck and short legs. (Both specimens remain on display in the American Museum.) His search continued until the end of August, when he shipped his finds to New York and began his own trip back to the East Coast. A brief visit to the family farm in Carbondale on the way fueled his ambition for another year in New York and hopefully another return to the field after two long, hot summers of disappointment.

Osborn, meanwhile, fell further into envy. It "has been a very costly season—with many disappointments," he wrote in a letter that September. If only 1902 would bring a turn in his luck.

Barnum Brown in 1919, four years after the first full *T. rex* specimen was revealed in New York City. He often joked that he lost his hair as a young man after encountering a mountain lion in a cave in Patagonia.

Henry Fairfield Osborn in 1890, shortly before he joined the fledgling American Museum of Natural History in New York. The son of a powerful railroad executive, he saw building the museum's collection of dinosaurs as a way to maintain the family prestige.

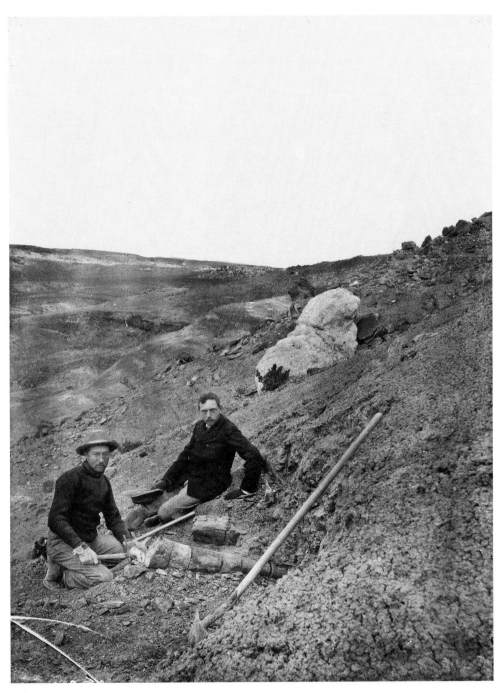

Well before they knew their lives would be integrally intertwined, Brown and Osborn met in the badlands of Wyoming when Brown uncovered the bones of a *Diplodocus*, the first dinosaur specimen in the American Museum's collection.

A view of the American Museum from the rooftop of the newly opened Dakota Building in 1880. The museum would eventually grow to encompass twenty-six interconnected buildings spanning across four city blocks.

Brown holds the reins as a team of horses hauls *T. rex* fossils weighing over 4,100 pounds out of a quarry in Hell Creek, Montana, in 1905. Brown discovered the first *T. rex* specimen in the same quarry three years earlier.

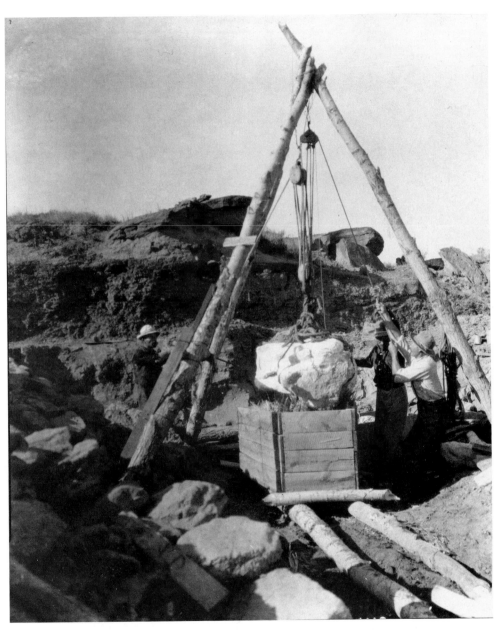

Brown (right) helps hoist the pelvis of a *T. rex* specimen he discovered in Montana into a crate to ship to the American Museum in 1908. The fossils are covered in plaster to protect them until they reach the museum, where the years-long process of mounting them onto an armature will begin.

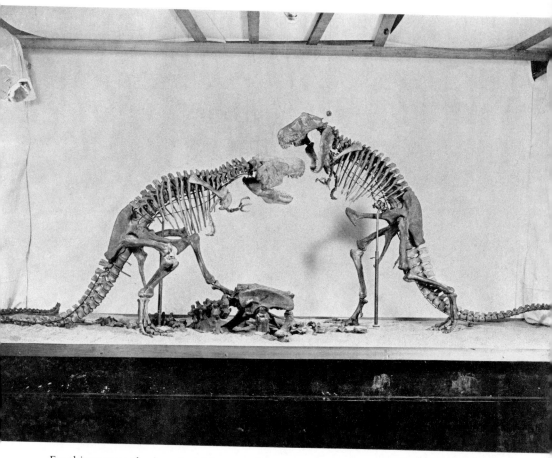

For thirty years, the American Museum of Natural History was the only place in the world where someone could view a *T. rex*. After Brown found multiple specimens, the museum briefly considered building an exhibit featuring two of the monsters fighting over prey, as depicted in this 1912 model, but scrapped it when it proved too large and unwieldy for the exhibition hall.

Despite severe nearsightedness that made him nearly blind, Charles R. Knight became one of the most influential museum artists of any era. This 1927 mural of a *Triceratops* facing off against a *T. rex* immersed the public in the prehistoric world and helped make *T. rex* the favorite villain of early Hollywood movies.

Brown, who grew up on a farm in Kansas, always hoped to become the cosmopolitan adventurer of his childhood dreams. Here he is shown wearing a fur coat while on a dig in Sweetwater, Montana, in 1914.

The *T. rex*, depicted here in its original "Godzilla pose" in 1960, soon became the most well-known dinosaur in the world and a fixture of everything from textbooks to breakfast cereals. In the early 1990s, staff at the museum repositioned the specimen in accordance with an updated scientific understanding so that its tail lifts off the ground and its head juts forward as if stalking its prey.

Chapter Eleven

THE BONES OF
THE KING

OSBORN SAT IN HIS OFFICE AS THE LOW WINTER SUN CAST
long shadows over Manhattan. Horses pulling buggies clomped
along the snow-dusted street outside his window, while men in
black bowler hats walked arm in arm through Central Park with
women ornamented by wide-brimmed hats crowned with feathers
and roses. From his chair, Osborn could look at a calendar reading January 1902, and wonder what the new year would bring.
He could faintly hear the sounds of construction coming from the
forest of upscale apartment buildings growing across the Upper
West Side in anticipation of the completion of the subway system,
two years away. The coming era of luxury was epitomized by the
twelve-story Dorilton, opening later that spring at the corner of
Seventy-First Street and Broadway, whose grand scale and lavish
ornamentation led the *Architectural Record* to call it an "aberration."
The building boasted separate servant and passenger elevators, fil-
tered water, and a parking space for owners who possessed an automobile, a novel invention which was only then starting to appear
on the city's streets. (One of Manhattan's first stand-alone "automobile stables"—a term that would soon give way to the modern
"garage"—opened on East Seventy-Fifth Street that year, prompting a magazine called *The Horseless Age* to marvel "we have never

heard of any [stables] being built solely for the use of automobiles.")
Further down Broadway, workers were putting the final touches
on the Flatiron Building rising above Madison Square Park, its tri-
angular frame prompting the photographer Alfred Stieglitz to say
that "it appeared to be moving towards me like the bow of a monster ocean steamer—a picture of a new America in the making."

History felt like it was speeding up, bringing in an era when
technology and science would dominate daily life. In the rush of
the new century, the traditions of the past seemed to fall one by
one to a new form of meritocracy built upon money, ushering in
a time when wit and chance mattered more than connections. It
was a time when the status enjoyed by the son of a railroad baron
suddenly felt diminished, as the churn of the economy made past
glories seem as inconsequential as a week-old newspaper.

While the world changed around him, Henry Osborn rested
his future on dinosaur fossils, intent on building a collection of
trophies that would stand as a monument to the millionaires who
funded the American Museum—and reflect some of their glories back on himself. Without dinosaurs, the museum was likely
to limp along in relative obscurity, becoming a relic as outdated
and ignored as some of the dusty items on its shelves. Fewer than
400,000 visitors—the majority of them schoolchildren, who did
not pay an entrance fee—walked through its doors that year, a
number equal to less than 10 percent of the city's population.
Unless Osborn could find the sort of gigantic fossil that would
draw paying crowds, the museum was on a path to a slow, stale
death, forever dependent on the whims of the very rich for its survival. "In the absence of any large income, we have been obliged
to depend upon the liberality of friends for the development of our
collections; and this will of necessity continue until our endowment is largely increased," its annual report warned that year.

Again, Osborn stretched for something to brag about to prove
his worth. "Progress has been made in rearranging the collections

so as to make them more intelligible to the public," he wrote in the annual report, straining for relevancy. "New specimens of interest have been placed in the center of the hall, and attention is called to them by explanatory labels, diagrams and models."

As to actual dinosaur fossils, the museum floor still held none. The partial *Diplodocus* that Brown had found was not yet fully prepared for exhibition. This process would repeat itself many times over Osborn's career at the museum: one of his collectors would bring in something momentous, yet the slow and exacting work of fully restoring the fossils and preparing them for a lifelike display would be drawn out over a span of years. Marsh, for all of his thirst for recognition, had found the chore of building armatures and restoring fossils too difficult to attempt—one reason, perhaps, why he disparaged those who mounted specimens in realistic poses. Osborn knew that the path forward was through such displays, yet he felt time ticking away with no guarantee that he would last long enough in his position to be there when the specimens were finally finished. In place of the bones still stuck in the back rooms and laboratories of the American Museum, Osborn convinced the trustees to purchase Cope's wide-ranging collection, which contained not only dinosaur fossils but those of prehistoric mammals and plant life as well.

When he felt cornered, Osborn tended to become more scathing and impetuous until he felt he had resumed his rightful place of superiority. Never a warm man, he turned colder as Hatcher continued to scorn his advances and poured salt in the wound with frequent friendly letters giving updates on his successes at the Carnegie. As Osborn planned for the summer prospecting season, he composed letters to his stable of collectors demanding that they ask more of themselves. "I hope you will push your work this summer with great energy and persistence, and accomplish fine results," he wrote to Granger. "Put as much heart into it as you can, because that is the direct road to success in everything; and I have had the feeling during this last year that you have not put quite enough of this element into your work." In a letter to Brown, he subtly highlighted the fact that, though he had been to Patagonia and back in the meantime, he had not found a significant dinosaur fossil since his first summer working with Wortman in Wyoming for the American Museum. "There is every reason to think that by careful inquiry among the natives, by making friends where you can, and by energetic prospecting, you may find something of real value," he wrote.

He knew Granger would take the slights without question; Brown, he recognized, was a wild card, driven by the same need for validation as himself. He hastened to add that should Brown find something, he should be prepared to stay quiet about it. "If you are striking it rich we will not say very much about it, because if we do we shall probably have some companion next year. I think that our friends in other Museums do not hesitate to poach on our preserves," he wrote.

He decided to send Brown once again on a mission for a *Triceratops*. The destination would be Montana, a state that was not yet known to hold abundant fossil beds. Most expeditions to date had centered on the proven fields of Wyoming, the Dakotas and Colorado, with each newly-discovered fossil bed usually lying within the same constellation as a previously-known site. Osborn directed Brown elsewhere largely on account of William T. Hornaday, the founding director of what is now known as the Bronx Zoo. Hornaday had explored Montana the previous summer with a photographer, intent on documenting the lives of blacktail deer in what would later be recognized as one of the first steps of the burgeoning conservation movement in America. In time, President Franklin D. Roosevelt would direct the National Park Service to name a mountain peak in Yellowstone National Park after him, in honor of his role in preserving the American bison from extinction. Yet, like other scientists of his generation, Hornaday could not reconcile his appreciation of the wondrous variety of the natural world with the fact that non-European humans lived in it, and remained unapologetically racist—once dismissing the controversy over his decision to include Ota Benga in the zoo's Monkey House with the words, "When the history of the Zoological Park is written, this incident will form its most amusing passage."

While traveling through Montana in 1901, Hornaday found himself in need of shelter and happened upon a friendly settler named Max Seiber, who offered him a room for the night. One evening became several, and Seiber showed Hornaday and his photographer around his vast ranch. He took Hornaday to "a spot nearby where he had found the badly weathered remains of what once had been a fossil skull, as large as the skull of a half-grown elephant . . . the skull was so badly weathered that nothing could be made of it, but near it lay several fragments of ribs in a fair state of preservation," Hornaday wrote. When he returned to New York, Hornaday showed the photographs of the find to Osborn and Brown, who instantly identified them as the remains of the elusive *Triceratops*. The only problem was that Hornaday could not remember where exactly Seiber lived, nor did he have any means of contacting him.

With a map of Montana and Hornaday's photographs to go by, Brown headed west, trying to find the location of the specimen and perhaps some of his lost self-confidence at the same time. Unlike the previous summer, which he had spent mostly in Arizona, he was once again on his own and free to follow his wits. He needed them. Montana stood as the largest haystack he had yet to search for a single fossil, an expanse stretching slightly more than 147,000 square miles over an area larger than Japan and nearly 60 percent larger than Great Britain. He knew that the *Triceratops* skull was there somewhere, and had the photographs to prove it. The hard part was retracing Hornaday's expedition with few clues to go by.

"It will not, however, be quite so easy to find it on the ground!" Hornaday warned in a letter that sat in Brown's bag as he left New York on a train that once again took him toward the unknown. "But you will manage that."

NOT SINCE PATAGONIA HAD BROWN seen a place as empty as this. Montana joined the Union as the forty-first state in 1889, and when Brown reached it thirteen years later it had more live-stock than people. Some six million sheep grazed on its plains, a population imbalance great enough to give each one of the state's human occupants a personal herd of twenty-five should they want it. Brown rode the train as far as Miles City, a small settlement that was the doorway to his ultimate goal: the Hell Creek Formation, a series of ragged red and gray ravines and stark badlands that were once a coastal plain but had dried out to become a sun-bleached wilderness. Though it had been sparsely explored by prospectors, the few relics collected from the region had proven to be of ex-traordinary size and quality. If Hornaday truly had come across a well-preserved *Triceratops* skull in Montana, then Hell Creek would be the most promising place to start.

Of all the inhospitable places that Brown's quest for dinosaurs had taken him, this was the most unforgiving. Hell itself would have been an improvement. The afternoon sun broiled everything in sight, while a haze of gritty dust perpetually hung in the air and choked the lungs of anyone foolish enough to try to breathe with-out a bandana covering their face. When it wasn't too hot and dry, it was too wet and windy. Sudden, violent thunderstorms boomed through the wide skies, unleashing a flurry of hail, tornadoes and floods that dared a person to survive. If the land didn't get you, then the animals would. Twisting, snarling ravines lurched across the badlands, hiding black bears, bison and rattlesnakes among outcroppings of rocks and fossils that date back more than 70 million years. In a land so barren, there was no hope of help if things went wrong.

The first known white explorers reached the region ninety-eight years before Brown. The Corps of Discovery, led by Meriwether Lewis and William Clark, stumbled into the badlands in 1804 while following the path of the Missouri River in the hope that it would lead them to the Pacific. The place unnerved them like few others they encountered on their long journey across the continent, haunted by a sense of menace that could not be explained merely by the realization that they were surrounded by untold numbers of dangerous animals. They wanted nothing more than to leave. "I sometimes wonder that some of our canoes or perogues [sic] are not swallowed up by means of these immence [sic] masses of earth which are eternally precipitating themselves into the river; we have had many hair breadth escapes from them but providence seems so to have ordered it that we have as yet sustained no loss in consequence of them," Lewis wrote in a journal entry on May 11, 1805.

When they could take a moment away from focusing on their own survival, they could not help but notice the strange rocks in the shape of bones that were jutting out of the earth. Officially, they were on an 8,000-mile expedition meant to find a waterway across the country, establish trade with the indigenous inhabitants and exert sovereignty over an area stretching from modern-day Louisiana to Montana, which the United States had purchased from France. Yet Thomas Jefferson, then the nation's president, requested that they also keep their eyes out for a mastodon, an elephantlike creature that scientists now know became extinct roughly ten thousand years ago but which Jefferson believed might still be roaming the far side of the continent. He had become intrigued by fossilized mastodon teeth discovered at a place known as Big Bone Lick in Kentucky, which Cuvier, the French naturalist, had highlighted as evidence that some species die out and disappear from the Earth. Though he had no religious objections to Cuvier's theory, Jefferson was not fully convinced that such a thing as extinction was possible in a rational, ordered universe. "In fine, the bones exist," he wrote. "Therefore the animal has existed. The movements of nature are in a never-ending circle." Following his curiosity, Jefferson turned to excavation and unearthed the remains of a "large lion-like creature" that he called *Megalonyx*—later identified as a giant ground sloth—in New Jersey and Virginia. No such thing still existed on the eastern seaboard; but then again, he reasoned, North America was a large place, and perhaps the *Megalonyx* and the mastodon had simply moved west before American colonists arrived.

Though the concept of a dinosaur did not yet exist, Lewis and Clark dutifully searched the landscape for intriguing stones or fossils and sent samples back to the White House. Jefferson soon opened a box containing a sampling of fish and mammal bones from Kentucky. Somewhere between present-day Sioux City, Iowa, and Omaha, Nebraska, Lewis and Clark came across "a petrified Jawbone of a fish or some other animal," a discovery which grew more perplexing as it became clear that they were not near an ocean. Indeed, the more desolate the region, the more likely it was to contain the bones of what looked to be aquatic life. In present-day South Dakota, they discovered the ribs, teeth and backbone of a creature that stretched 45 feet long. Not knowing what else it could be, they deemed it a large fish; scientists now believe that the specimen was likely the remains of a plesiosaur, a marine reptile that lived alongside dinosaurs in the Mesozoic Era.

The Corps of Discovery were likely the first white explorers to find dinosaur fossils in the western half of the United States, though they did not know it. On their way back from the Pacific, they stopped at a 150-foot-tall block of rock they named Pompey's Tower, in honor of the eighteen-month-old son of Sacagawea, a Lemhi Shoshone woman born in present-day Idaho who acted as a guide and interpreter for the Corps as it traveled across the continent. On the same day that he carved his name into the rock—now known as Pompey's Pillar—Clark noted in a journal entry full of creative spellings that he "employed himself in getting pieces of the rib of a fish which was Semented within the face of the rock this rib is about 3 inchs in Secumpherance about the middle. . . . I have Several peces of this rib the bone is neither decayed nor petrified but very rotten. The part which I could not get out may be Seen, it is about 6 or 7 Miles below Pompey's Tower . . . about 20 feet above the water." Perhaps with the memory of a whale carcass he had viewed on the Pacific Coast fresh in his mind, Clark reasoned that the bones were those of a similar sea creature worn away over the centuries. Modern-day paleontologists believe that he was describing a *Hadrosaurus*, a duckbilled dinosaur that is among the more common specimens in the area.

By the time Brown reached the Hell Creek region nine decades later, dinosaur fossils were becoming a well-known and increasingly profitable part of the landscape, though it had yet to see the same sort of frenzied activity as the fossil beds of Wyoming. "Yesterday I heard of another specimen having been found within five miles of Forsyth," he wrote in a letter to Osborn in June 1902, shortly after he reached Miles City. "I rode out [to find] . . . the skeleton having been destroyed by souvenir hunters. It was a *Claosaurus*"—the same species of duckbilled herbivore that Brown had found the previous season without much excitement. Its skull and vertebrae were "sold to the Smithsonian Institute [sic] about four years ago. It seems a Baptist Sunday School teacher had stolen the skull and sold it," he added.

Brown continued his search and within a matter of weeks managed to find the ranch where Hornaday had snapped pictures of what appeared to be a *Triceratops* skull. Upon close inspection of the bones, however, he realized that it was not a *Triceratops* at all, but the broken remains of a mosasaur, a giant predatory marine reptile that, like modern snakes, had two rows of teeth in its upper jaw to prevent prey from escaping. Fearing that another summer would end with him returning empty-handed, Brown resolved to stay in Hell Creek until he found something that would ensure his future with the museum. He bought three horses, a wagon full of equipment and enough food to feed a small expedition party. He was soon joined by Richard Swann Lull, a doctoral student of Osborn's at Columbia University, and a young volunteer named Phillip Brooks.

Tall, athletic and pompous, Lull was six years older than Brown and outranked him in terms of academic achievement, yet Brown had more field experience and knew how to survive in the harshest conditions. Still, Brown did not wear his higher status well. It was his first time heading a group expedition for the museum, and the unfamiliar managerial tasks weighed on him after the freedom of solitude he had enjoyed in Patagonia and Colorado. Even little things tripped him up. He wrote constant updates to Osborn but often neglected to send them, confessing in one dispatch that he had "discovered the letter still in my pocket yesterday so I will try a new one." There was so much to do, and so many things that could go wrong—not only for him, but for the museum, which desperately needed some good news from the field.

In early July, Brown's party arrived in the hamlet of Jordan, a hundred miles northwest of Miles City, a roughly five-day wagon ride that felt far greater. Founded three years earlier by a hunter named Arthur Jordan, it had the grim distinction of being located farther from a railroad station than any other settlement in the country. Its lone civic building was a post office built out of logs; its most frequented was a saloon that kept its beer in the cellar, providing the only cold drink within a hundred miles. In such a lawless, isolated and dangerous place, the arrival of men from New York City was a diversion that could not go unrecognized. "Every shady character that could not stand the spotlight of civilization drifted in and around the new town, always ready to have fun or start trouble, and a few of them desired to pick on the scientists and their men," Arthur Jordan wrote in his autobiography. "Only once did they annoy Professor Lull when he came in town to get his mail. . . . They ran for their rifles and began to throw lead all about him. The fellow made his team do their best in going over the hill and away from those drunken, demented morons."

Brown kept the expedition team moving before the citizens of Jordan had a chance to improve their aim or regain their sobriety, and in the following days they made camp on the banks of Hell Creek. He took a horse and went on solitary scouting trips through the ravines and foothills, trying to get a sense of how the land was shaped and where promising layers of stone might lie. Brown had a talent for surviving in nearly any conditions, but was still taken aback by the difficulty of the terrain. "The country greatly resembles Lance Creek in Wyoming along the breaks but the main canyons are certainly *bad* lands, almost impossible lands I might say," he wrote in a letter.

Even so, he could not let up. In one of the most hostile places on Earth, the demands of New York found a way to reach him. A letter from Osborn arrived on July 25, admonishing him for the wasted freight charges and labor after a shipment of fossils he had collected earlier that summer arrived at the American Museum in a broken heap due to the rough transport. "You will be very much disappointed to learn that the Dinosaur which you collected with so much care and labor has proved almost valueless. . . . It will perhaps yield two or three bones of value," Osborn wrote. "The skull proves to be entirely crushed and unrecognizable. This seems to warn us that we should certainly examine material a little more carefully in the field. . . . I know you sent the specimen to us after the best possible methods; but it should have received a more careful examination."

As the last days of July ticked away, Brown found himself

torn. The party uncovered a *Triceratops* skull that was in decent condition, though its horns were missing. With enough work, it could be "a fine exhibition specimen," he wrote to Osborn, knowing that would begin to make up for the crushed fossil now sitting in the museum in New York. If anything, the skull would buy him at least one more year of employment. But he wanted more. Never one to be satisfied with what he had in hand, Brown felt compelled to brave another ravine, search another hillside, climb over the side of another cliff if doing so meant that he would come closer to a specimen that would put the trajectory of his life back on its upward tracks. As the temperature soared above 100 degrees, Brown worked without stopping, the contours of his face slowly disappearing behind accumulated layers of grime and dust. Since childhood he had acted as if he had an unspoken trust that the universe would bend in his favor when he needed it. Each morning he stepped out of his tent seemed to be another plea that his luck would return.

As July faded into August, Brown attacked a sandstone hill he called Sheba Mountain with a plow and scraper, determined to satisfy his curiosity about what lay beneath. Its particular composition of stone and its location near what was once an inland sea fit the profile of a promising fossil bed. Once Brown began to dig, however, the rock proved incredibly hard, seemingly impervious to any blade. Unable to let it alone, he sent an assistant to Miles City to come back with enough dynamite to blow off all the hillside above what he hoped would be the bone layer. Brown was not in the habit of blasting away at every spot that gave him trouble, but this time—whether due to frustration, intrigue or a combination of the two—he had to know what secret the Earth was protecting with such ferocity. He laid the explosives, set the timer and waited. The blast echoed among the ravines of the badlands, reverberating like distant thunder. A dark cloud of dust and dirt hung in the air, so thick that he could taste sand on his tongue. Once the smoke cleared, he edged closer to the lip of the quarry, staring into the deep hole he had created. It was nothing less than a time machine, bridging the 60-million-year gap between the age of the dinosaurs and our own. As he looked down into the pit, Brown took in a shape that no human being had ever laid eyes on. "Quarry No. 1 contains the femur, pubes, [partial] humerus, three vertebrae, and two undetermined bones of a large Carnivorous Dinosaur not described by Marsh," Brown wrote in a letter that evening to Osborn. "I have never seen anything like it."

It was as if a child's conception of a monster had become real and was laid down in stone. Though most of its skull and tail were missing, everything about the beast seemed designed to overwhelm the human mind: its hips, nearly 13 feet above the ground, would later be found to power legs that ran at speeds greater than 10 miles per hour; its immense jaws measured over four feet in length and could exert as much pressure as the weight of three modern cars, instantly exploding the bones of its prey; its serrated teeth, the longest of any known dinosaur, could dig through the thick skin of a *Triceratops* and rip out five hundred pounds of flesh in one bite. In time, the creature would become perhaps the most recognizable animal the world has ever seen, its deadly silhouette and Latin name familiar even to those with no interest in dinosaurs or science. Yet in that moment in the hot August sun, the animal that would soon take the name of *Tyrannosaurus rex* was entirely new—an unmistakable set of clues that the history of life was more varied and surreal than anyone had imagined.

Brown knew that he was suddenly in a race against time. He had found the only specimen of a creature previously unknown to science, and there was no telling if he or anyone else would ever find another. With less than two months before the broiling landscape would become too cold to allow work to continue, Brown scrambled to uncover as much of the fossil as possible. A September snowstorm was not out of the question, which could mean that the dinosaur might have to be abandoned over the long winter and spring—enough time that someone from Jordan could pick up on rumors of its discovery and try to sell it themselves, or, even worse, destroy it through carelessness.

Brown rode his men hard, his natural cool reserve vanishing under the demands of an unrelenting taskmaster. The small general store in Jordan soon ran out of lumber and plaster, the two supplies most essential in extracting a gigantic fossil out of the ground without damaging it. With no other choice, Brown turned to dynamite, taking the risk that the surrounding rock layer was dense enough that he would not accidentally blow up the fossil before he could share it with the world. "The bones are separated by two or three feet of soft sand usually and each bone is surrounded by the hardest blue sandstone I ever tried to work in the form of concretions," he wrote to Osborn in September, after nearly a month of nonstop digging and prying. Each day, more of the animal was revealed, like the wrapping paper of a gift being removed inch by inch.

Finally, in October, Brown pulled the last section of the skeleton free. The small team of horses strained under the load, pulling the haul to Miles City in shifts, eventually moving more than fifteen thousand pounds of bones to a boxcar that Osborn had arranged for them. As the first snow of winter fell, Brown watched as nineteen crates of fossils were loaded into the boxcar, a collection that included not only the new carnivorous dinosaur, but also the skeletons of a crocodile-like *Champsosaurus* and a *Triceratops*— both of which still stand on the floor of the American Museum.

Though he knew that its contents were priceless, he did not accompany the train east. A prospector at heart, he could not leave Montana without poking his head around for at least a few days more. He soon walked into the lobby of the Billings State Bank on a small errand and, while waiting in line, noticed a display case containing the oversized limb bone of a dinosaur. A brief talk with the bank's manager revealed where the fossil had come from, and Brown spent the final days of the year camped on the banks of Beavers Creek searching for the remainder of the specimen.

When the train carrying the crates from Jordan reached New York, Osborn did not realize what he had. It would be two years before museum technicians fully cleaned and prepared the skeleton of the *T. rex*, removing each piece from the matrix of stone and arranging them all in a fully articulated skeleton. Until then, what in time would become the most famous dinosaur in the world looked no different from every other fossil Brown had found, encased in a cocoon of white plaster with the letters AMNH painted in black on its side.

With nothing to quell the sense that he was losing the fossil race, Osborn remained preoccupied with the Carnegie Museum and its *Diplodocus*, unable to see the magnitude of what Brown had discovered. "I think we have the finest carnivorous Dinosaur material in the world; but I envy the Carnegie Museum their complete skeletons. Mr. Hatcher writes me that they have found a magnificent *Diplodocus* which seems to be almost perfect," he wrote in a letter at the end of the prospecting season. "We are certainly not holding our own . . . both Carnegie and Chicago have done better than we have."

Chapter Twelve

NEW BEGINNINGS

NOTHING ABOUT THE CREATURE MADE SENSE.

In a laboratory high above the American Museum's exhibition halls, Lull directed a team of preparators as they cut open the plaster jacket holding the bones of the newly-discovered dinosaur and began chipping and chiseling at stone that seemed unnaturally strong. Mice scurried under their feet as days passed without noticeable progress. Still they worked on, taking care not to let their impatience get the better of them. In the field, the work of a paleontologist was all action, locked in a battle with the elements to free a specimen from the earth before it crumbled; in the laboratory, the work was painful and slow, a monastic discipline that required delicacy above all. A collector had to trust that another dinosaur fossil always lay around the bend; a preparator never forgot how close they were to disaster. Every action was a hypothesis put to an instant test. One wrong choice—pressing too hard, sanding too closely, brushing too vigorously—and a bone more than 60 million years old could be shattered, destroying priceless evidence of Earth's history that might never be found again. Once all traces of rock and dust were finally cleared from a section of bone, it was coated in layers of shellac to prevent it from crumbling. Work then turned to the next segment, rebuilding the puzzle of life piece by painful piece.

Over the course of two years, Brown's discovery began to take shape. Osborn watched as the creature's jaws, vertebrae, ribs, shoulder and pelvic bones emerged from a bed of rock, all the while imagining how they fit into a living, breathing animal. The questions the specimen posed came faster than any answers. First among them, why did an immense creature have such strangely small and seemingly useless forelimbs? Nothing as diminutive had been found in the fossil record among carnivorous dinosaurs, nor in any of their prey. Osborn stretched for an evolutionary explanation, eventually landing on the idea that perhaps the two-fingered arms were used for "grasping during copulation."

Thinking about how exactly dinosaurs had sex was not out of the ordinary. The question remains a pressing line of inquiry in paleontology today, given that a dinosaur's sexual anatomy would indicate how closely it resembled modern-day birds—which pass semen through the cloaca—or whether it was more reptilian in nature and had a penis, like modern-day crocodiles and alligators. Regardless of anatomy, the act of mating would seem to pose a challenge given the need to navigate long and bulky tails and, in species such as *Stegosaurus*, the presence of spikes that would suggest a risk of injury or castration. As recently as 2013, one computer-assisted model suggested that the only way to solve the problem would be for a female *Stegosaurus* to lie on her side, though no one knows how long her partner's penis or cloaca could stretch, because soft tissue rarely fossilizes.

Still, Osborn had his doubts that the needs of mating were sufficient to explain the diminished anatomy of Brown's new discovery. While the size of the forelimbs "abundantly characterizes this animal . . . in the writer's opinion final judgment must be suspended until the skeleton is fully worked out," he wrote when formally describing the specimen. In a small footnote below a sketch of the full skeleton, he made it even more plain. "The [positioning] of the small forearm is probably incorrect," he wrote. The forelimbs were not the only puzzling aspect of the beast's bones. Upon inspecting them, Osborn realized that its hind limbs were hollow, like those of a bird. With this, the creature became the most prominent example of how prehistoric life blurred the lines of the animal kingdom.

Some forty years before Brown's discovery, workers at a limestone quarry in Solnhofen, Germany, unearthed a nearly complete fossil marked by clear impressions of spread wings and tail feathers etched in the stone. Its anatomy suggested two creatures at once: the feathers and the air sacs in its backbone indicated a prehistoric bird, while the body—including developed teeth, a long tail and three clawed fingers capable of independent movement, unlike the fused fingers of living birds—pointed toward a reptile. (Some who examined the specimen, which cast a beautiful white outline against the yellow tint of rock, argued for a third option: fossilized angel.) A local natural history professor determined in 1861 that it was a reptile. The British Museum of Natural History soon purchased the find, and Richard Owen decided to reclassify it as a bird with the name *Archaeopteryx macrura*—all but ignoring the evidence that the specimen was just the sort of transitional fossil between two species that Darwin had theorized in the *Origin of Species* but had yet to be found in the fossil record. Owen was widely derided by his fellow English naturalists, who were disinclined to like him anyway, with one contemporary writing in a snide letter to a friend that the specimen was "a much more astounding creature than has entered into the conception of the describer." Not to be outdone, Darwin wondered, "Has God demented Owen, as a punishment for his crimes, that he should overlook such a point?"

Darwin cited the fossil—which became known as the "London specimen"—in subsequent editions of his landmark book as an example of his contention that life does not have fixed categories. "Hardly any recent discovery shows more forcibly than this, how little we as yet know of the former inhabitants of the world," he wrote. At roughly 160 million years old, *Archaeopteryx* is now considered one of the earliest known birds. Studies of its DNA show that it was covered in jet-black feathers, like a raven. Nor was it alone in its likely coexistence with dinosaurs; in 2005, paleontologists in Antarctica discovered a 68-million-year-old creature, known as *Vegavis iaai*, which looked remarkably like a modern duck, suggesting that other birds were there in some form all along. *Vegavis iaai* most likely quacked like a duck, too, given that it is the oldest known example of an animal with a vocal organ called a syrinx, which allows modern birds to make their distinctive sounds. No dinosaur fossils with a similar structure have yet to be uncovered, suggesting that, though many of them were likely feathered, dinosaurs did not sing.

A small animal that combined features of reptiles and birds was one thing; a massive carnivorous dinosaur with architecture similar to that of a creature capable of flight was quite another. As he studied the specimen in front of him, Osborn slowly began to realize that its importance lay beyond its shocking size. It was a scientific marvel in every sense, unlocking a door to the distant Earth that had been closed for 60 million years. Never before had such a large predator been found, much less conceived of. Its existence implied a complex ecosystem far removed from the land of fat, lazy giants that Owen had imagined. Gigantic herbivores were well known at this point, signaling the existence of a lush prehistoric environment that provided enough plant life to sustain them. A carnivore of the same size, on the other hand, suggested that animal life must have been more plentiful and dense than previously imagined. Otherwise, how could a predator of such an enormous size find enough food to survive? And if life was more plentiful, that implied at least some form of social structure, suggesting that dinosaurs could have been capable of moving in herds, like landbound flocks of birds.

The questions posed by Brown's discovery came in waves. A creature this huge and ferocious had to hunt, but how? It had to roam, since otherwise it would conceivably exhaust all sources of food quickly—but how far? And if an animal like this existed, its potential prey must have developed some form of defense—but what? Nothing before had ever demonstrated the ferocity of evolution in such an obvious form; nothing else made the idea of slow but potent change so visceral. At the end of the late Jurassic period, an extinction event erased giant sauropods like the *Brachiosaurus* from the Earth. Some 40 million years later, life had somehow found a way to reorder itself into the monster whose bones were now spread out in a Manhattan laboratory.

Its size struck Osborn for other reasons as well. Throughout his career, he had a tendency to view the history of life on Earth as a sort of morality tale in which good prevailed in the end. In that light, Brown's discovery was the perfect example of the shortcomings of physical strength: huge, brutish and extinct, an unmistakable illustration that muscle alone does not guarantee survival. Its disappearance opened up the ecological space for mammals, which were blessed with greater intelligence and the capacity to care for one another. The failure of a species with a body so powerful, and the triumph of humans with bodies so weak by comparison, seemed to Osborn evidence of a great plan, a long pageant of life that peaked with present-day Anglo-Saxon humans in their rightful position of power.

Osborn had long wanted a giant that would prove his worth; now, he had the biggest meat-eater that ever lived. By 1905, he understood enough about the animal to let the world in on the secret. The first person to publish a scientific description of a newly discovered species gets the honor of naming it, following an international convention which typically incorporates some combination of a description of its distinctive features, where the species was found or the person who found it. *Stegosaurus*, for instance, has its root in the Greek word for roof, inspired by Marsh's ultimately incorrect assumption that its plates lay flat on its back as a sort of protective barrier above its body. Other well-known species were named for their distinctive horns (*Triceratops*), the sound they likely made when they walked (*Brontosaurus*, the thunder lizard) or the simple fact that they hadn't been seen before (*Allosaurus*, the strange lizard). Hatcher, with his decision to name the *Diplodocus* specimen after Carnegie, had expanded the scope of possibilities to include patrons of museums, infusing bones many millions of years old with the power to reflect a human's social worth.

Hatcher upstaged him once; Osborn was now in a position to return the favor. In a move that broke with convention and displayed an element of showmanship reminiscent of P. T. Barnum, Osborn announced that the species that Brown had unearthed would be named *Tyrannosaurus rex*, meaning tyrant lizard king. The animal was "the ne plus ultra of the evolution of the large carnivorous dinosaurs: in brief it is entitled to the royal and high-sounding group name which I have applied to it," he wrote. The name was lyrical—its parade of r's sounding almost like a chant before the snakelike resonance of the "ex"—and it was justified, a reflection of the creature's outsized body and position at the top of the prehistoric food chain. But beyond the scientific reasons he could point to for bestowing on the creature such a distinctive and alluring name, it spoke to Osborn's dreams of glory. He would forever be linked with the one species known as a king, elevating his own place in paleontology above any competitors. With this creature, his time had finally come.

Though *T. rex* now had a name, no one outside the museum had yet seen it. In February of that year, the American Museum unveiled a 70-foot-long *Brontosaurus* (now known as *Apatosaurus*) as the centerpiece of its new Dinosaur Hall, a combination of metal rods, screws and plumbing equipment holding the fossilized bones together in a lifelike pose. It was the first time that a fully assembled sauropod had stood on the floor of the museum. Thousands of New Yorkers came to view the giant, intent on seeing the beast worthy of a private reception hosted by J. P. Morgan a few days before. Few knew what exactly to call it, the word "dinosaur" proving to still be a stumbling block among those who had little exposure to science. "Some wanted to see the 'dino', others the 'diorso' and among other destinations were 'the octopus' and 'His Nibs, Old Boney.' The attendants saved embarrassment by announcing, before a question could be framed, 'fourth floor to your right,'" noted the *New York Times* in a small article on the opening of the Dinosaur Hall.

The reaction of visitors once in front of the bones seemed a mirror into their own lives. One boy asked an attendant if the dinosaur would eat him. (Don't worry, he wouldn't eat "little fellers" like you, he was told—just plants.) A butcher calculated how many pounds of meat he could reap from the animal. A truck driver pondered aloud what kind of traction the beast's claws gave it on a slippery road. It was as if ordinary New Yorkers stumbled upon a passage into a prehistoric world, only to find themselves there.

For many who stood in front of the *Brontosaurus*, it was the first time they were confronted with unmissable evidence that the world of the past did not look like the world of today. The presence of the dinosaur was like a window shade pulled open, revealing that science was not an abstract concept meant only for the privileged but something that could be as universal and understood as sunshine. The presence of working-class visitors in the museum did not sit well with everyone. A "professor with large glasses" who visited the exhibit called it "in bad taste in a place devoted to science," the *Times* wrote. His companion, however, was of a different mind. "'I don't like it but I must excuse it,'" he said, according to the *Times*. "'It has drawn all these people here, being a splendid bit of advertising. They are, many of them, deeply interested in things they would have never come to see. They will come again, and bring others.'"

Attendance rose 25 percent that year, a considerable jump for an institution nearly forty years old. Yet while the new dinosaur specimens brought in new visitors, few lingered in the hall. The bones of Jumbo, the circus elephant, remained the most popular exhibit. A few weeks after the *Brontosaurus* was unveiled, the usual crowds once again formed around the gem exhibits and other favorites, leaving the paleontology rooms sparse except for those who had yet to see its main attraction. Osborn, though he did not want to accept it, could not help but notice that one impressive dinosaur was simply not enough. For paleontology to be more than a season's novelty, the museum would have to move past displays of isolated specimens and show the vast complexity of the former world, effectively turning dinosaurs from curiosities into the manifestation of a powerful idea. A mounted *T. rex* would provide a natural counterbalance to a huge sauropod, implying a vision of a reordered world that could not be found in any one-off display of a P. T. Barnum–type novelty.

There was a problem, however. As the preparators grew closer to finishing their work extracting the specimen from the matrix of stone, Osborn could not look past the fact that it was lacking too many key components for the museum to build a realistic display. Three years after Brown discovered the first *T. rex*, Osborn sent him back into the unforgiving landscape of Hell Creek on a mission to find another one. This time, however, he wasn't alone.

✛ ✛ ✛ ✛

BROWN NEVER STAYED IN ONE place for long. He could be found anywhere but New York in the summers since he had uncovered the *T. rex*, often toiling in the sort of unforgiving places that people went out of their way to avoid. Yet every fall he would return, and over that cycle of arrival and departure he had somehow found a way to bring Marion back into his life. She became a constant for him, a stable tether to the present that was often lacking in his profession. She often ended her days teaching at Erasmus Hall in Brooklyn, then the city's most prestigious public high school for science, to find Brown standing outside waiting for her. The two spent hours walking through Brooklyn's Prospect Park, building a connection based on their shared love of the natural world. She pointed out the various species of birds and insects they passed, granting Brown the novel experience of learning about living creatures, rather than about those that were long dead. In Marion, Brown had for the first time found someone whose combination of intellect and humor allowed him to drop his defenses and feel comfortable, and ignore his instinct to roam. She, in turn, felt a thrilling sense of freedom when she was with him, and delighted in his distinctive mix of purpose and fun.

They decided to make it official. On February 13, 1904— one day after Brown's thirty-first birthday—they were married at St. Paul's Episcopal Church in Marion's hometown of Oxford, New York. The day was cold and sunny. While waiting before the ceremony, Marion wore bright red socks over her white satin slippers to keep her feet warm. She had already started walking down the aisle on her father's arm when she realized she still had them on. "With a muffled giggle and two vigorous kicks, she got rid of them just in time. Not that Barnum would have cared if she appeared at the altar in red socks," their daughter, Frances, later wrote. "Many friends of both the bride and groom often spoke later of Marion's blonde, almost ethereal beauty as a shaft of sunlight through a stained-glass window struck her happy face."

For their honeymoon, they spent five months in the field prospecting, finally knitting together the two sides of Brown's life. In Marion, Indiana, they inspected a nearly complete mammoth skeleton found by local amateur collectors. From there, they headed for the badlands of South Dakota, where Marion kept detailed notes of the local birds while Barnum searched for a plesiosaur, a type of long-necked marine reptile. It was her first taste of the frontier, so different from the cultured confines of home. Like Brown, she delighted in the humor of everyday life and latched on to the unintentional comedy that came from mixing scientists focused on prehistoric wonders with ranchers whose only experience in the world was survival. One afternoon, Brown and a field assistant explained to a local rancher that they were searching for the bones of a mosasaur, a predatory marine reptile that could reach up to 50 feet long. The rancher calmly replied, "Yes, I see one of them darned things swimming down in Hat Creek the first year I was here," Marion wrote in an unpublished memoir of their trip she called *Log Book of the Bug Hunters*.

Marion quickly proved that she belonged. One night, the expedition camped near a stream fed by meltwater, filled with speckled trout. Marion improvised a hook from a safety pin and baited it with grasshoppers. She "caught several messes of good-sized beauties," Brown later recalled. "They were a welcome change from canned food." Over the following weeks, they braved every extreme the West had to offer: "high pinnacles, cathedral spires, and fortresses of fantastic shapes"; temperatures which climbed to 105 degrees; hailstorms accompanied by more than four inches of rain within an hour; and vast swarms of mosquitoes "too numerous to mention," Marion noted. Like Brown, she never let the obstacles of nature block her attempts to understand the world around her. While he worked to excavate a duckbill skeleton near the Judith River, she took detailed notes while watching a toad give birth to thirteen babies. "The old toad never paid the slightest attention to them after they were born," she wrote.

They took a circular route across the country. In Colorado, they rode the railway to the top of Pike's Peak. In New Mexico, Marion studied the techniques with which Navajo women wove patterned woolen blankets while Brown searched nearby rock outcrops, eventually discovering a new specimen of hadrosaur now known as *Kritosaurus navajovius*, which is currently on display in the American Museum. From there, they traveled to Arizona and then to Arkansas, becoming more alike with every passing day. One night in camp, Brown spied a copperhead snake poised to strike Marion. He whispered to her to remain perfectly still while he took out his revolver and killed it with one shot. "Decades later, Barnum remarked that Marion's instinctive reaction to his unexplained command was an example of the perfect rapport between them," their daughter later wrote. On a swing back through Arizona, they camped on the edge of a thousand-foot cliff high above the Grand Canyon. When they returned to New York a month later, Marion set up a small tank in her classroom and plopped a horned toad she'd found in Arizona inside, a living souvenir of their journey.

The following summer, Osborn directed Brown to return to Montana and search for another *T. rex* to complement the specimen now nearing completion in the museum's laboratory. He urged Brown to concentrate on acquiring enough material for a display mount, rather than filling up the museum with the bones of other, less important animals. "Every portion of [a *T. rex*] will be valuable," he wrote. Brown readily accepted the assignment. When he left for Montana that summer, Marion was once again at his side, having proven her ability to handle the field over their long honeymoon the year before. Brown left the arrangement unspoken, neglecting to bring it to Osborn's attention yet taking no pains to hide it. For him, Marion had become another part of himself, as inseparable from his body as his nose.

The expedition returned to the quarry near Jordan, Montana, in 1905. In the three years since Brown's discovery of the *T. rex*, what was once a dangerous and isolated place had become a regular stop on the summer prospecting circuit. Brown arrived back in Jordan to learn that a crew from the Carnegie Museum had swept through a few weeks before without finding any worthy specimens, though they had respected the custom among collectors and left his previous quarry alone. Local ranchers, meanwhile, were now aware that the bones that Brown and the scientists from the East Coast spent so much time looking for must be valuable, and grew hesitant to let paleontologists explore their land without a contract that would give them an ownership stake in any discoveries. According to federal law, any fossil found on private land is the property of the landowner, while specimens found on public land are the property of the U.S. government. Thanks to dinosaurs, the purposeless violence found in Jordan had evolved into the mania of a gold rush, a mixture equal parts desperation and greed. One family, known as the Sensibas, offered to sell a specimen Brown identified as a hadrosaur, which they had found on their land, for $8,000, or roughly $250,000 in today's money. Brown, who knew a farmer's bluff when he saw it from a childhood watching his father sell coal, cautioned Osborn to limit any contact with the Sensibas so as not to make the museum appear anxious and become a mark for others attempting to unload bones at inflated prices.

No one but Brown had ever found a *T. rex*, leaving open the question of whether it was a feat that could be repeated. Hoping that his luck would not run out, Brown led the expedition team back to his previous spot in the badlands. He soon located the quarry and began the process of reopening it, blasting away at the rocks he had used to seal it and building new earthworks to support further excavation. He found something almost immediately. Less than six feet away from the location of the specimen he had discovered in 1902, he uncovered a *T. rex* skull that when pulled from the stone weighed more than eight hundred pounds. The job of fully extracting it, however, proved to be one of the greatest physical challenges Brown ever faced. The job "will take at least three weeks with powder and horse power for it is a solid sand bank," he wrote to Osborn, adding a request for a half dozen short, heavy chisels of the "best steel." One block of stone containing bones from the specimen weighed 4,150 pounds and required six horses to pull it out of the quarry. Round after round of blasting and scraping expanded the pit to 100 feet long, 20 feet deep and 15 feet wide, by far the biggest that Brown ever made. "This is a heavy piece of work but [*Tyrannosaurus*] bones are so rare that it is worth the work," he wrote to Osborn in late July.

As he dug deeper, Brown kept careful note of the rock layers he unveiled. Near the end of the season he felt confident enough to share with Osborn a theory that he had been formulating for several seasons. "I am fully convinced after several years of work that the [coal-bearing] beds are separate and distinct from the [dinosaur-bearing beds]. I have yet to find a dinosaur bone in the [coal-bearing beds]," he wrote. Over the next two years he continued to refine his thoughts and eventually published one of the rare scientific articles of which he was the main author. In the paper, he described the features of what are now known as the Hell Creek and Tullock formations. Over time, the rock layers that caught Brown's eye would be found to have high levels of iridium, an extremely dense metal that is rare on Earth but unusually abundant on asteroids and comets. The presence of such material from a likely interstellar source is now seen as powerful evidence for the theory that a large asteroid or comet played a key role in the extinction of dinosaurs by drastically altering the Earth's climate.

In early August, Osborn surprised Brown with word that he planned to travel to Jordan to oversee the last stages of the dig, his impatience getting the better of him. Only then did Brown confess that Marion had been at his side the whole summer. "Mrs. Brown did accompany me from my home to camp and had done the cooking for the outfit this summer reducing our living expenses about half," Brown wrote, hoping to appeal to Osborn's sense of thrift. "I did not discuss the matter with you for it seemed a purely personal matter with me as long as I performed my duty without any added expense on her account and the Museum has certainly been the gainer."

Osborn ultimately did not venture out to Montana that summer and made no mention of Marion's presence on the expedition team. Perhaps in another summer, or with another collector, he would have. Yet the relationship between the two men had changed. Brown had proven himself at the time of Osborn's greatest need, and, should he ever tire of Osborn's ways like other collectors had before him, he now had the distinction of having discovered the world's most fearsome carnivore—a résumé item that would be sure to draw the attention of the Carnegie or any other institution that wanted to rival the American Museum. For the first time in his career, Osborn was forced to treat Brown with a light touch, granting him deference that nothing in Brown's background or social class would otherwise merit. Though his profile had risen with the jump in the museum's attendance following the opening of the *Brontosaurus* exhibit and his association with the *T. rex*, Osborn had not yet climbed as high as he hoped. He still needed Brown—and his ability to find impressive specimens—as a bridge to becoming the king he envisioned himself to be.

The world's most famous dinosaur at the time was far from his hands. In 1902, King Edward VII visited Andrew Carnegie in his castle in Scotland and, upon seeing a sketch of the skeleton of a *Diplodocus* hanging on the wall, asked what it could possibly be. "The hugest quadruped that ever walked the Earth, a namesake of mine," Carnegie replied. Unfamiliar with the difficulty of finding a complete set of fossils, the king asked if Carnegie had a similar specimen that he could spare. Carnegie offered a plaster copy, and in early 1905 thirty-six boxes containing 292 replica bones arrived in London.

Four months later, two hundred men and women crowded into the Natural History Museum's Gallery of Reptiles in South Kensington for the formal unveiling of the replica, the first time that any form of the *Diplodocus* specimen had been assembled for public viewing. (The original bones remained in Pittsburgh, and owing to the time needed to build supports that could withstand their massive weight would not be ready for display until 1907.) Newspapers from around the world were on hand to record the scene. Edwin Ray Lankester, the museum's director and a man who was two years away from being knighted for his contributions to British science, strode to the stage. Once uncharitably described as "a massive yet nimble mind dwelt in a massive frame," Lankester was not yet ready to hand over the crown of paleontology to the country that was the source of its largest specimens. "All the great progress that has been made in the American Republic has been founded upon ideas, which have germinated, and inventions, which have been really conceived, in England," he said, adding that the *Diplodocus* was simply "an improved and enlarged form of an English creature."

Carnegie let the infighting pass, looking toward something greater. He was one of the first to grasp the true power that gigantic dinosaurs—whether in the form of models or reconstructed skeletons—had on the public imagination, fostering a sense of unity through an appreciation of the size and mystery of these extinct beasts. The great cost and labor put into the dinosaur replica was successful because through it "an alliance for peace seems to have been affected . . . jointly weaving a new tie, another link binding in closer embrace the mother and the child lands," Carnegie said in a speech at the replica's unveiling in London.

The gift to the Natural History Museum was the beginning of what became known as Carnegie's dinosaur diplomacy, and he soon gave additional replicas of the *Diplodocus* to museums throughout Europe in the belief that forging cultural bonds between countries would prevent future wars. The London specimen, now known as Dippy, became a national phenomenon. Visitors packed into the hall, leading a board member of the Natural History Museum to speculate that one day's attendance might have been higher than its draw over the previous twenty years.

Osborn had nothing that garnered such worldwide attention. As the 1905 field season drew to a close, he pinned his hopes on the future success of the *T. rex*, hoping that the monster he held in storage would vault him—and the institution—past his rivals.

"The results of our season elsewhere have been very disappointing," Osborn confided in a letter to Brown as the summer ended.

Chapter Thirteen

THE HARDEST WORK
HE COULD FIND

NEW YORK IN 1906 WAS A CITY BURSTING WITH DIVER-
sions. At the Polo Grounds, the New York Giants drew nearly half
a million people to watch a season spent in the distant wake of
the first-place Chicago Cubs, who ended the year twenty games
ahead in the standings. In Brooklyn's Park Slope neighborhood,
the Superbas (soon to be renamed the Dodgers) attracted nearly
300,000 spectators that season despite the fact that they were not
very good, ending the year seventh out of the eight teams in the
National League. Putting that many fans in the seats was itself
something of a miracle, given that the stadium was right next to
the pungent Gowanus Canal, a grimy waterway so polluted by
industrial waste that a state commission called it "a disgrace to
Brooklyn." For those looking for indoor forms of entertainment,
the Metropolitan Museum of Art that year brought in just under
800,000 visitors, many of them lured by a decision by Sir Caspar
Purdon Clarke, the museum's director, to move the collection of
modern sculptures away from the walls. The decision "has over-
come the effect of emptiness hitherto presented by this part of the
Museum," the *New York Times* wrote in appreciation. "This not
only improves the appearance of the hall, but enables visitors to
study the statues from all sides."

Getting tourists to actually come through the doors of any museum was itself becoming a problem. After starting the city's first roving tour guides with a team of six horses and an open carriage a few years earlier, an entrepreneur from Colorado named Henry J. Mayham improved his operation with the purchase of a multilevel, open-topped motorized bus. A driver sat at an oversized steering wheel while to his right a guide with a megaphone shouted out factoids of history. The tour rumbled along a twelve-mile route, giving visitors a way to experience the city while hardly setting foot in it.

Not that you needed to go out into the streets for excitement. Millions of New Yorkers grabbed their newspapers each day to read the latest in a murder trial gripping the country. Harry Thaw, a Pittsburgh millionaire and husband of Evelyn Nesbit—the most famous model in the country—had shot Stanford White, whose architecture firm of McKim, Mead and White had transformed New York over a generation through the design of buildings ranging from Madison Square Garden to the Washington Square Arch, three times in the head on the evening of June 25, 1906, as he sat in the Garden watching a musical number entitled "I Could Love A Million Girls." After Thaw was apprehended that night, he said he did it because White had raped his wife when she was sixteen and he would kill him again if given the chance. The upcoming legal case would soon be called "the trial of the century" and mark the first time that a crush of media coverage prompted a judge to sequester a jury in American history.

World-famous art, professional sports, high-society murders: in such a cityscape, it was hard for anything—even a 40-foot-long *Brontosaurus*—to stand out. As the year wound to a close, Osborn counted the daily visitor tallies at the American Museum of Natural History, which painfully confirmed a truth that his own eyes had told him for months. Attendance was down nearly 15 percent from the year before. In that time, he had been promoted to president of the museum, putting the full weight of the institution's future on his shoulders. Though it was a scientific marvel, the *Brontosaurus* exhibit was not enough of a draw on its own to make the museum a destination among the multitudes of options in New York. Osborn's expectation that mounted dinosaur displays would continue to draw crowds looked increasingly foolish, as if he were simply dressing up the hokum and novelties of P. T. Barnum with the pretensions of the upper class.

He had no other choice but to push more chips onto the table, hoping that the problem was that his bet wasn't large enough to begin with. In a final effort to put a good face on a bad year, he unveiled an exhibit featuring the feet, legs and lower pelvis of the *Tyrannosaurus rex* on December 29, 1906. It was the first time any portion of the creature had been put on public display, and broke with Osborn's reluctance to display any dinosaur, much less one as important as the *T. rex*, before the full mount was ready. Yet he felt boxed into a corner. He desperately needed some good press for the museum, if only to convince donors that their money was not wasted.

A few days before the exhibit opened, the *New York Times* described the *T. rex* as "The Prize Fighter of Antiquity Discovered and Restored" in an article that incorrectly bolstered Osborn's role in finding the specimen. "Of the tyrannosaurus, the greatest of flesh-eating animals, the only known specimen is the one discovered in Montana by Prof. Henry F. Osborn, and now mounted and placed on exhibition for the first time in the Natural History Museum," the paper reported. A photo that ran alongside the piece showed a museum worker whose height reached slightly below the creature's knees, dusting its pelvis with a broom. To a modern eye, nothing about the scene is recognizable as the *T. rex*. Instead, it looks like a spindly pair of stone legs, waiting for a body to appear.

In the frenzy to share the species with the world, Osborn neglected a fact that had been obvious since the first dinosaur fossils were put on public display: visitors might be awed by the size of an animal's limbs, but what they really want to see is its head. That fact was what compelled him to send Barnum Brown into the field searching for the skull of a *Triceratops* over several summers until he found one worthy of display, and it was the skull of the museum's mounted *Brontosaurus*—rather than its dramatic tail or long neck—that visitors were drawn to, as if looking for a personal connection between themselves and the deep past. In time, the skull of the *T. rex*—with its oversized nostrils, imposing size and sinister jaws and teeth—would become one of the most recognizable silhouettes in all of science, an emblem of everything from children's pajamas to amusement park rides. But at that moment Osborn had nothing to show. Brown, the only person yet to find a *T. rex*, continued to search for a skull worthy of exhibition after the previous season's find proved incomplete. No one knew if he would ever unearth one, leaving Osborn in the awkward position of promising that more was coming, like a painter describing the plan of a mural while he waits for his brushes to arrive. The legs alone of the *T. rex* did nothing to excite the public imagination and offered no hints to the fantastical predatory machine the creature once was. Instead, the exhibit seemed as if it was nothing more than bones unconnected to a larger idea, a lone piece of a puzzle far separated from the box it came in.

Osborn's usual solution to any problem was to send Brown into the wilderness and expect him to come back with something important, once again living up to his nickname "Mr. Bones." Yet for the first time in nearly a decade, that option was unavailable. Brown remained in Brooklyn with Marion throughout her pregnancy and was at her side when their daughter, Frances, was born on January 2, 1908. He began the adventure of fatherhood in an apartment very different from the rural farmhouse in which he grew up.

The break from digging gave Brown his first sustained opportunity to focus on the academic side of paleontology, finally freeing him to turn to theory after having proven his worth in the field. He spent months working out the details of how to mount the museum's collection of *Anatosaurus* specimens—a large duck-billed dinosaur that lived in North America roughly 65 million years ago, which was distinguished by rows of hundreds of blunt teeth along its elongated jaw. He also wrote some of the few professional papers he published in his lifetime, detailing not only the sediment layers of the Hell Creek Formation but the abundance of fossil mammals he uncovered in a 1903 expedition to northern Arkansas known as the Conrad Fissure Collection, including black bears, wolves and more than fifteen different species of what are now known as saber-toothed cats.

Though he tried, Brown was not built to stay in one place for long. The world was too big, there were too many things to pursue and too many places to explore. Becoming a father did little to make it easier to ignore the voice that forever told him to seek out new places, in rebellion against the confines of his childhood. After Frances was born, he could no longer resist it. With Marion's blessing, he began planning an expedition to the hard landscape of the Big Dry region of Montana, about thirty miles east of the Hell Creek beds where he discovered the first *T. rex*. The sediment formations were similar, increasing the chances that another important specimen could be found. If anything, they were more remote than those near Jordan. It was a part of the same pattern of discovery that marked his career; each major find would draw the attention of other paleontologists and museums, forcing Brown to go farther and farther into the blank spots of the map in search of something new.

With Marion and Frances nesting in the family apartment in Brooklyn, Brown left New York in early June with a renewed appreciation of the freedom and possibilities of the open West. Within three weeks he was drinking and singing late into the night at raucous dinners with local ranch families and making plans for an even bigger celebration on the Fourth of July. After months of city life and the responsibilities of parenthood, he grabbed at any opportunity for fun, convincing a local woman to drive around the county collecting phonographs in order to throw a party. "I danced till twelve finding some very good partners and one bad one," he wrote in his field journal. "Some of the people danced all night till seven o'clock Sunday morning then had breakfast and drove to their homes."

Science seemed almost an afterthought in those wild weeks. Yet it was his counterintuitive ability to find fun that made Brown such a successful collector. In terms of technique, he was unmatched: he seemed to have an innate feel for the land, reading the rocks and pushing past the physical constraints of others in ways that he could not teach. But getting to that point—knowing where to dig a new quarry, or how to have the luck to find a dinosaur fossil jutting out of the ground—was often a reflection of his mastery of the softer science of friendship. With his deadpan wit and willingness to try anything at least once, Brown innately knew how to get people to like him. More often than not in the lonesome communities that bordered the badlands, those conversations would lead to invitations to come out and inspect a strange bone that happened to be sitting on a rancher's land. What to an outsider seemed like a charmed life of continuous discovery was the result of Brown's tending to both poles of his personality: the intellect and drive which made him chafe at what he saw as the dead end of Carbondale, and the joy he took in the presence of other people that often put him at the center of any party. The results of this two-pronged life in the field often made it seem like Brown had a touch of magic and was fated to find fossils where others were cursed to come home empty-handed. "Found Tyrannosaurus lower jaw and back of skull near one of the buttes. Will take it," Brown wrote in his journal in early July.

Not long after, he found fifteen connected tail vertebrae from an animal he could not immediately identify. The layers of sediment were relatively soft compared to the conditions in nearby Hell Creek, making it possible for him to dig an additional six feet into the rock without needing to turn to dynamite. The line of bones continued, suggesting the presence of a complete skeleton. He borrowed a plow from a local rancher in order to remove layers of topsoil and rock, and, over the following two weeks, he and an assistant chiseled the bones free. Finally, on July 15, he wrote to Osborn with news of another discovery. "Our new animal turns out to be a Tyrannosaurus," he wrote. "The bones are in a good state of preservation."

That was not all. The new specimen was more complete than any that Brown had yet found, and included the most elusive prize: a perfectly preserved skull. For the first time, human eyes took in the complete four-foot-long head of a *T. rex*, confirming in one glance the violence that it was once capable of. Osborn immediately made plans to travel west, to see the dig that he knew would change paleontology—and his career—forever. "Your letter of July 15, makes me feel like a prophet and the son of a prophet, as I felt that you would surely find a Tyrannosaurus this season," he wrote. "I congratulate you with all my heart on this splendid discovery. . . . I am keeping very quiet about this discovery because I do not want to see a rush into the country where you are working."

The discovery of a fully intact *Tyrannosaurus* presented Brown with problems beyond the necessity of keeping it a secret. First among them was getting the bones out of the ground without disturbing them. Brown fired the expedition's cook—no small gesture in a place so remote that finding a replacement was not guaranteed—after he caught him driving a team of horses into camp falling-down drunk. "Was sorry to lose him for he was a good cook and a good worker but I won't have a man in camp that I cannot trust to [drive to] town with the bones. Hope we may have someone who can cook without burning the water when you come out," he wrote in a letter to Osborn in early August. The caution was warranted. As Brown removed more of the rock surrounding the specimen, he began to realize just what he had. "I am sure you will be more pleased with our new Tyrannosaurus when you see what a magnificent specimen it is. The skull alone is worth the summer's work for it is perfect," he wrote to Osborn in mid-August, noting that the lower jaws of the creature weighed about 1,000 pounds.

Osborn arrived on August 26 and attention immediately turned to the question of how to get the bones back to New York safely. Brown, who still felt the sting of Osborn's rebuke when a shipment of unrelated fossils had arrived at the Museum destroyed, insisted on loading the full specimen onto one train rather than parceling it out into several shipments to save on costs. Osborn relented, leaving Brown in a race against the coming winter. Snow started falling in late September, well before he was finished building the crates to hold the collection of bones now encased in plaster. Still he worked, dragging and digging and dynamiting until he pulled each section of fossil from the rock. The loads were so heavy that wagons kept breaking down under their weight. Twice Brown had to abandon the entire specimen in the Montana badlands when the wagons pulling it got stuck in a rut, while he rode ahead to a local ranch to borrow more horses. In the end, it took sixteen horses to pull five loads of fossils forty-five miles to the nearest rail depot, finally bringing the full splendor of the *T. rex* out of the dusty pockets of the badlands and into the modern world.

THE *T. REX* WAS A revolutionary find; to do it justice, Osborn felt an obligation to reimagine how a dinosaur appeared in a museum setting. With the partial mount's failure to attract attention the year before still haunting him, the question of how to present a full specimen to the public consumed Osborn. For all of the clues they offered about Earth's past, dinosaurs had not yet proven any power to hold the public's attention for a sustained amount of time, and seemed to be exhibits grounded more in a sense of spectacle than in science. The same problem had hung over paleontology since the field's inception: a suspicion that it was a pursuit of the shocking and grotesque rather than a legitimate field of study that offered some connection or insights into modern-day life. In the public's mind, dinosaurs were little more than one of Carnegie's replicas of the *Diplodocus* or a *Brontosaurus*—huge animals whose main value was their sheer size, without any deeper meaning.

The *T. rex* exhibit had to do more. Osborn wanted to convey a sense of movement and danger, making visitors feel that they were stepping onto a planet that looked very different from today's— a proposition that subtly implied that the world might also look very different tomorrow. He asked an artist working in the paleontology department to sculpt a scale model of the animal's full skeleton, and then asked several designers to submit their best ideas. Eventually, he chose a display suggested by Raymond Ditmars, who had helped found the Bronx Zoo's popular reptile house with more than a dozen species from his own collection and would go on to produce groundbreaking natural history films. Instead of a staid pose like Carnegie's *Diplodocus*, Ditmars proposed a scene in which a *T. rex* is in the middle of devouring a duck-billed herbivore when another *T. rex* attempts to steal its meal. "The crouching figure reluctantly stops eating and accepts the challenge, partly rising to spring on its adversary," Brown wrote in a paper describing the planned exhibition design. "The psychological moment of tense inertia before the combat was chosen to best show positions of the limbs and bodies, as well as to picture an incident in the life history of these giant reptiles." The size of the exhibition hall would ultimately prove too small to contain Ditmar's proposed mount, however, leaving Osborn to scale back his ambitions.

Over the course of his career, Brown discovered dozens of species and rarely concerned himself with how their final displays would appear. Yet the *T. rex* was different. In its monumental size, Brown saw a validation of his own ambitions. He had worked his way out from his family farm and discovered a creature that would stand forever in a prominent spot in what was becoming, thanks to his efforts, one of the most important museums in the world. It was a culmination of all the lonely nights spent in the middle of nowhere, wondering if his luck would continue and lead to another fossil or finally run away from him, the last gasp of borrowed time. He had seen the world, married a woman he loved and become a father, fulfilling all the dreams that he dared to ask of himself. And with a perfect *T. rex* skull still undergoing its slow restoration process in the museum's laboratories, he knew that there were greater accolades to come.

OVER THE NEXT TWO YEARS, Brown returned to Hell Creek and ventured farther north into Canada, always on the lookout for new relics of the prehistoric past. He uncovered a *Triceratops* skull, fossilized crustaceans and a crocodile–like *Champsosaurus*, continuing to fill back rooms in the museum while preparators were busy readying his previous discoveries for exhibition. But he always returned to Brooklyn, his life falling into a dependable pattern: summers in the field and the rest of the year in the city, making a daily circuit between the museum and the family apartment. After his childhood on the frontier, he had successfully built a bridge into a new era, making himself relevant in a changing world in ways that his own father could not.

With its skyscrapers, subways and palaces of culture, the New York of 1910 looked in many ways like the city of today. Yet it was a city still rooted in the past, beset by the illnesses of urban life that had plagued cities since the Middle Ages. Measles, rheumatic fever, diphtheria and whooping cough killed thousands of New Yorkers that year. The pain was felt heaviest by children. Nearly sixteen thousand newborns in 1909 did not live long enough to celebrate their first birthdays, a death toll nearly triple that of any other city in the country. Nor were their mothers spared. What was then known as the puerperal state and is now commonly called post-partum was responsible for 45 percent of all deaths among adults between the ages of twenty and twenty-nine, making the act of giving birth nearly three times as deadly as getting into an accident and twice as likely to be the cause of death as suicide.

While the cities that it compared itself to seemed to be getting healthier, New York fell in the other direction, with the death rate among children under five at the start of 1910 higher than that of London or Paris in the 1880s. To be a parent in 1910 was to face each day with fear that a wayward cough or sneeze in a crowded streetcar might unleash an invisible killer that would attack the most vulnerable. Each year that a child lived seemed like another battle won. The war would continue until their fifth birthday, the age at which the diseases of urban life seemed to lose much of their lethality and allow parents to finally trust that their offspring would not be taken from them too soon. Scarlet fever was among the city's most notorious killers. Named for the bright red rash that covers the body, it was caused by the same bacteria that causes strep throat. Children who caught the disease would suffer with a high fever and would find it painful to swallow, their tongues often swollen, with a white coating, and a pale ring appearing around their mouths.

Barnum and Marion Brown celebrated when Frances turned one, thankful not only for the joy she had brought into their lives but for the fact that she had not been taken from them. At the age of two, she remained strong and healthy, showing some of the same grit that allowed her father to outlast his competitors in the field. That strength was tempered by her mother's calm and measured disposition. Barnum "told his daughter more than once that it was her mother, Marion, who had the brilliant mind, not her father. He agreed that his was an average mind but that Marion's was far better," Frances later wrote.

On a sunny day in April 1910, Marion placed Frances in a stroller and took her on a walk through nearby Prospect Park. Though Marion would later not recall seeing anyone who was sick, Frances soon spiked a fever and broke out in a bright red rash. Within a day, Marion developed the same symptoms. Their fevers rose and rose again, unbroken by any remedy. As they spiraled into sickness, Barnum did not know where to direct his fears. Each time Frances seemed stable, Marion would get worse; when Marion felt calm, Frances would appear closer to death. Antibiotics to treat scarlet fever would not be developed until 1928, nearly a full generation in the future. Instead, Barnum was left to frantically search for anyone who could help save his wife and their only child.

Frances's fever soon broke, but Marion continued to get worse. The family's doctors were at a loss, unable to extinguish the fire that raged inside her body. Brown finally asked Osborn for help, hoping that his wealth and connections could do what medicine so far had not. Osborn immediately began making calls, promising Brown that he would bring in specialists from anywhere in the world at the museum's expense. But it was too late. Five days after entering Prospect Park as the healthy mother of a smiling young girl, Marion died on April 9, 1910.

In Marion, Brown had found the only person who had tamed his innate sense of chaos. With her passing, his tether to a normal life was slashed. The question of how to live seemed overwhelming; the duty of fatherhood simply too much. "Marion's shocked and grief-stricken parents, both then in their middle sixties, told Barnum that they would take the baby and raise her, that they could not have anyone else have all that was left to them of Marion. Barnum, torn with grief and anger at cruel Fate, agreed that that would be the best solution for the daughter," Frances later wrote in a memoir of her father's life. "Long afterwards, as a mature woman, Frances could understand how her father reasoned in that desperate time: a daughter was expendable; a wife was not."

A light went out in Barnum Brown's life. In shock, he clung to routine and formality as best as he could. "I want to thank you for the efforts to help keep my beloved wife alive, your kind letters of sympathy, and financial aid—all expressions of truest friendship for which I cannot find words to express my appreciation," he wrote in a letter to Osborn. "I shall always be deeply grateful. The baby continues to improve but still shows kidney complications. . . . The funeral service will be Tuesday, May 17th, at 4 o'clock and I shall return to my duties Wednesday."

Throughout his life, Brown chased the horizon to solve his problems. Less than a month after losing Marion and giving Frances to her grandparents to raise, he set off for a hastily-planned expedition into the isolated Canadian wilderness. It was there, Frances later wrote, that he was "free to seek numbness from pain by throwing himself into the hardest work he could find."

Chapter Fourteen

A New World

It was as if he wanted the elements to kill him.

The Red Deer River Valley is an isolated and unforgiving region about 130 miles east of Calgary, Canada. Gray chimneys of stone known as hoodoos soar over the rippled badlands, which rise and fall in an unsettled rhythm. Rocky mesas give way to steep canyons that plummet as if a trapdoor has been pulled. It is a land marked by absence. Valleys where fecund groves of palm trees and fir once flourished, millions of years ago, are now carved by dry river beds called coulees, which trace below a hard landscape of cliffs painted in bands of colors ranging from sand to copper to a gray so deep it appears purple. Squat bushels of muted green scrub brush layered with thorns dot the ridges and valley floor, undisturbed by any trails or paths. Aside from the shimmering water of the river, the only burst of color comes from pockets of sunflowers growing out of stone.

During their journey across the continent, Lewis and Clark attempted to explore the length of the Red Deer River, but turned back after finding that a series of violent rapids obstructed any passable routes. Subsequent wars between the Cree and Blackfoot nations stripped the area of most of its human population, and by the turn of the twentieth century only a handful of white ranchers braved a place where in the summer great swarms of mosquitoes attacked without mercy and in the winter violent blizzards blotted out all avenues of escape. Few roads existed as late as 1910, giving the region a timeless quality that made it impossible to tell what year, or century, it was.

There, in one of the most secluded spots in North America, Barnum Brown attempted to escape the weight of his pain. Racing to get as far away from New York, Frances and the memory of Marion as he could, he packed enough supplies to last for months adrift from the wider world. He set off without much of a plan in mind, almost as if he were counting on running out of luck after a lifetime of close calls. He built a flat-bottomed boat and pitched a tent on its deck, creating a mobile camp that ensured isolation. River currents took him through canyons with walls looming more than 250 feet above. The emptiness of the land was healing, his sorrow drowned out by the scale of the natural world. "No more interesting or instructive journey has ever been taken by [this] writer," he later wrote. "Habitations are rarely discernible from the river, and for miles one travels through picturesque solitude unbroken save by the roar of the rapids."

The area would not have been known to science had it not been for a surveying party that passed through some twenty-six years earlier led by Joseph B. Tyrrell, an arts graduate from the University of Toronto who turned toward geology after his physician advised him to work outdoors to help him fully recover from a childhood bout of scarlet fever. He took a job with the Canadian Geological Survey and was assigned the task of exploring the unmapped western wilderness. He covered more than 116,000 square miles of the province of Alberta on foot, canoe and dog-sled before reaching the Red Deer River Valley in search of coal deposits. On June 9, 1884, he uncovered the first known dinosaur fossils in the region. Further investigation over the following days revealed that the area was teeming with bones, as if he had walked into an immense dinosaur graveyard.

Tyrrell did not linger on his discovery, however, and the Red Deer River Valley remained undisturbed until Thomas C. Weston, a Canadian geologist, convinced a local rancher to build a boat and float fossils down the river. The teamwork paid off and in 1889 Weston found the skull of what is now known as an *Albertosaurus*, a carnivorous dinosaur that lived approximately 100 million years ago and may have evolved into the larger *Tyrannosaurus* group. The small size of Weston's boat limited the number of bones he could collect, however, leaving the region largely untapped as paleontologists and museums scoured similar landscapes in Montana and Wyoming. Though Osborn and Brown knew of reports that abundant fossils had been located there, the Red Deer River Valley only became a point of focus in 1909, when a rancher from the area visited the American Museum of Natural History and asked to speak with a curator. He was soon sitting in an office on an upper floor, giving Brown detailed descriptions of the large fossil bones jutting out from the canyon walls on his ranch, which looked just like those in the museum halls below.

Brown possessed some of the few maps in existence that marked likely fossil sites, but during those lonesome weeks on the river chance was his guide. He had long been known for his willingness to work harder than anyone else, refusing to leave picked-over fossil beds until he spotted something that everyone else had missed; now, in a landscape where there were no second chances, he lost himself in work, letting his world shrink down to the size of the next bend in the river. The boat rode along the current until Brown spotted a potential prospecting site and guided it ashore. "In the long midsummer days, at latitude 52 degrees, there were many hours of daylight, and constant floating would have carried [the boat] many miles per day; but frequent stops were made to prospect," he wrote, resulting in an expedition that "rarely covered more than twenty miles per day."

The routine was soothing, a daily reminder that he had a purpose. He could not control the unseen virus that had robbed him of the woman who once served as the anchor amid the chaos of his life, nor could he face the fact that he felt unfit to serve as the only remaining parent in his young daughter's life. With nothing left, he did the only thing he knew how to do. Over two months of searching, Brown uncovered fossils, dozens of them, all of them worthy of display. He remained in the field deep into the fall, refusing to turn back as the first week of September bombarded him with sheets of rain and sleet. "The fossils are so numerous that I doubt if we can possibly work farther down the river than Tolman's [Ferry] this season," he wrote to Osborn. "This is without doubt the richest Cretaceous deposit in America . . . with our numerous boxes on board containing such a variety of creatures we are living in a veritable ark."

He relented as the full force of winter made it impossible to continue on. Over the next few months, he never stayed in one place for long. After a quick visit to the family farm in Carbondale, followed by a short trip to Oxford, New York, to see Frances, he raced south to Texas and then to Mexico, where he discovered fossilized mammoths. By March 1911 he was in Mississippi, where he uncovered the remains of primitive whales; by the start of April he was in Florida, staying just over a week before sailing on to Cuba. Once there, he explored caves and mineral springs to help determine whether land bridges once connected Cuba and other Caribbean islands to the mainland of Florida or Mexico, going so far as to build a machine to drain a hot spring so that he could better excavate its bed. He uncovered the remains of alligators, crocodiles, turtles and a nearly complete skeleton of a *Megalocnus*, a giant sloth that went extinct just a few thousand years ago, which he shipped back to New York. Soon it stood in the museum's Hall of Mammals and Their Early Relatives.

Yet wherever he traveled, his thoughts kept returning to the secluded Red Deer River Valley, a place whose emptiness allowed him to let go of his pain. Two weeks after returning from Cuba, he once again headed north to Alberta, where he arrived in mid-July. There, he felt at peace, and little by little allowed himself to enjoy the small pleasures of life for the first time since Marion's death the year before. "I picked and ate 4 different kinds of berries— strawberries, gooseberries, raspberries, and saskatoons," he noted in his field journal one night. He stayed all summer and returned the next, coming back to a land filled with distractions.

He slowly felt the joy of collecting coming back. With it came a renewed sense of daring. Soon, he was hanging off the side of a bluff, dangling by a rope as he chiseled away at rocks that he hoped contained fossils. A few weeks later, he led a field crew farther down the river on a flat-bottomed boat, their bodies completely covered by thick clothing despite the broiling heat. "Work has gone rapidly but under trying conditions," Brown wrote. "I cannot approximate the number of mosquitoes but every person who moves about is forced to wear a net over the face, gloves, and a coat or extra heavy shirt. I have never experienced anything like it."

When Marion was alive, Brown remained faithful to her out of both love and duty, taming his inclination to seek new adventures and new company. As he continued on his self-imposed exile from New York and the memories of her, he once again felt free to follow his instincts, regardless of the consequences. Other members of the expedition began to vent in their notes about Brown's frequent overnight absences, often spent at the home of a local lumberman, Roy Hard, when Hard's work had taken him out of town. For the rest of his life, tales of Brown's womanizing would follow his expeditions into the field, with whispers among staff members at the American Museum of repeated payoffs to settle paternity suits. (Brown's numerous love affairs were so well known that, upon news of his death, an assistant at the museum rushed into his office

and destroyed folders full of his personal letters and correspondence, depriving scholars of the details of his many entanglements.)

Brown's success in the field soon brought unwanted public
attention. As details of his finds leaked beyond the small circle
of the American Museum, Canadian newspapers and scientists
started to ask why an American was allowed to export some of the
country's greatest treasures to New York without paying for them.
The field of paleontology in Canada was at the time a shadow of
that of its southern neighbor, having no version of a Bone Wars or
competition among millionaire-backed museums to prod it forward. Hoping to close that gap, the Geological Survey of Canada
asked Charles H. Sternberg and his three sons to collect specimens
that would remain in Canadian hands.

In many ways, Sternberg was a glimpse into what Brown's life
would have looked like had he missed his chance encounter with
the American Museum while in college. The son of a minister,
Sternberg had spent part of his boyhood on a farm in Kansas, where
he filled the long hours collecting fossils that lay scattered around
him just waiting to be picked up. When he was seventeen, he
sent specimens of plants fossilized in sandstone to the Smithsonian
Institution, which replied with letters encouraging him to send
more material. He turned a season of collecting for Cope in western Kansas while studying at Kansas State University into a career
as a freelance collector, eventually selling dinosaur specimens to
major museums ranging from San Diego to Sweden, as well as the
American Museum in New York. Each summer, he traveled to
the most remote areas of North America, surviving on hard tack,
beans and wild berries while prospecting for fossils. Every winter,
he cleaned and prepared his discoveries for sale and built a public
following by writing articles for general interest magazines, a book
of poetry and several memoirs, the first of which, titled *The Life of
a Fossil Hunter,* was published in 1909.

A deeply religious man, Sternberg was animated in part by the thrill of uncovering previously unseen parts of what he saw as God's master plan. He viewed his purpose as a link in the chain, connecting holy evidence of the prehistoric past with those who would live long after his own time on Earth was over. "I shall perish, but my fossils will last as long as the museums that have secured them," he once wrote. "My own body will crumble in dust, my soul return to the God who gave it, but the works of His hands, those animals of other days, will give joy and pleasure to generations yet unborn."

Brown learned of Sternberg's plan to prospect through the Red Deer River Valley through his son, George, who was working with the American Museum that summer. "Papa and the boys will soon come up in this part of the country for the Ottawa Museum," George wrote in a letter early in the summer of 1912. The appearance of another prospecting party in a region to which Brown retreated for solace awoke a part of him that had laid dormant since Marion's death, and he seemed reanimated by the prospect of competition. "The Ottawa party are somewhere . . . twenty miles below which does not disturb me," Brown wrote to the head of the museum's Department of Vertebrate Paleontology. "As long as they are there I shall concentrate the whole party in this formation where the exposures are best."

In early September, with only a few weeks remaining in the collecting season, Brown set off in a motorboat 150 miles downstream, to reach a rock outcropping that he had noticed the year before while on a surveying mission. There he found and began excavating the nearly complete skeleton of a previously unknown duck-billed herbivore with a crescent-shaped helmet on its skull. Brown dubbed it *Corythosaurus*, meaning Corinthian helmet lizard, yet he was struck by something else about the specimen: its underside was covered in skin impressions, making it appear more like a mummy than a collection of bones. He had never seen anything like it. For its skin to have remained intact, the body of the animal must have fallen into water with an extremely low level of oxygen and then been quickly buried by sediment, protecting it from both predators and microbes. Brown rushed to excavate the specimen before exposure to the sun and the weather destroyed it, and shipped it back to New York just as winter made further explorations impossible.

Over three summers in the Canadian wilderness, Brown rebuilt himself, shedding his grief and replacing it with a renewed desire to explore. In the darkest moments of his life, the sense of restlessness that had sustained him as a child returned to save him, turning the process of collecting fossils into a declaration that he would not succumb to hopelessness. As he steered the boat containing more than twenty cases of fossils toward the railway station that would take him and his discoveries back to New York, he lamented in a letter that the riverbanks were so full of specimens that he had to "travel with my eyes shut so as not to see more that I want this fall."

IT TOOK BROWN DAYS TO recalibrate when he returned to New York. After a summer removed from nearly all the elements of modern life, the packed-in city seemed overwhelming, a crush of people unimaginable when floating down a Canadian river just a few weeks earlier. Brown often spent nearly all of his first days back in the city inside the museum, huddled with Osborn and the members of the Department of Vertebrate Paleontology briefing them on what he had found and checking on the progress of the mounts of his discoveries from past seasons in the field. He would then walk the exhibition floors, noting how visitors approached the fossils he had uncovered and which drew the largest crowds. That there were finally crowds, at all, was apparent, a reflection of Osborn's efforts to bring in more tours of schoolchildren in hopes that they would return with their parents in tow. The museum's

attendance jumped 17 percent over the previous year, edging closer to the one-million-annual-visitors mark, "although the location
of the Museum is still far from the center of population," Osborn
noted in the annual report.

Brown began noticing a strange pattern: each year when he
returned from the field, dinosaurs seemed to be more popular than
they had been before he left. The wing holding the partial *T. rex*
exhibit was now among the most visited in the museum, drawing
attention away from the bones of Jumbo the elephant, which had
delighted a generation of children before. Every year, the collective memory of P. T. Barnum's museum of novelties receded a
little further into the past, clearing mental space for dinosaurs to
be seen as scientific specimens untainted by showmanship, finally
perceived as ideas rather than distractions. Since his college years,
Brown had spent his life trying to fill the halls of the museum;
now, as he neared forty, that work had changed the world, making
it seem entirely natural that extinct monsters would be displayed
for any child who wished to see them.

THE INCURSION OF DINOSAURS INTO everyday life was a phenomenon that went beyond the halls of the American Museum.
In 1912, Sir Arthur Conan Doyle took a break from writing the
Sherlock Holmes novels that he despised, despite their popularity,
and created a character who was the opposite of his most famous
creation. Where Holmes was cerebral, melancholy and deliber-
ate, Conan Doyle's new Professor Challenger was loud, physical
and impulsive, a man whom his wife calls a "perfectly impossible
person." It seemed only fitting to send him to South America,
where Challenger leads an excursion to prove that he has discovered prehistoric life in the Amazon. There, the party finds a nest of
pterodactyls, which Conan Doyle described as repulsive monsters.
"There were hundreds of them congregated within view. All the

bottom area round the water-edge was alive with their young ones, and with hideous mothers brooding upon their leathery, yellowish eggs. From this crawling flapping mass of obscene reptilian life came the shocking clamor which filled the air and the mephitic, horrible, musty odor which turned us sick," he wrote.

When published in 1912, *The Lost World* was a global success, marking one of the first times that dinosaurs played a central role in a work of popular fiction. Its resonance with readers rested not only on the thrilling depictions of prehistoric beasts battling modern-day humans, but also on what dinosaurs were starting to represent: tangible, irrefutable evidence of not only evolution, but a path of evolution which had its endpoint in Anglo-Saxon civilization. Conan Doyle's depictions of Africans who only wish to serve their European masters, of "savages" and "half-breeds" and apemen, reflected a widespread belief that race was intrinsically linked to intelligence and ability. This line of thought ran through the American Museum in displays ranging from African artifacts to a diorama depicting the first encounters between English settlers and Native Americans in New York. As president of the museum, Osborn installed exhibits that implied an upward ascent of evolution culminating in what he called the Nordic race. The Hall of the Age of Man, for instance, began with the displays of the ancient peoples of Africa before moving to the tribes of North America and ending with Europe, suggesting a ladder in which light-skinned humans were the product of billions of years of refinement.

The twin fields of natural history and conservation were filled with men who saw no separation between their beliefs in science and in racial hierarchy. The endpoint of this line of thought was the eugenics movement, which held that applying the principles of genetic selection to the human population would cure the Earth of "undesirables," ranging from the physically infirm to darker-skinned peoples. Madison Grant, a fellow member of the Manhattan aristocracy who helped found the Bronx Zoo and created the first organizations dedicated to protecting the California redwoods and American bison, argued that humans had now replaced natural selection, granting the powerful "complete mastery of the globe" and "the responsibility of saying what forms of life shall be preserved." In 1916, Grant would write a polemic titled *The Passing of the Great Race, or The Racial Basis of European History,* in which he argued that noble Nordic instincts and talent for self-governance were being diluted by the population growth of Alpine and Mediterranean peoples. Adolf Hitler wrote a letter to Grant calling the book "my bible." In a foreword to the book, Osborn praised Grant's theories by writing, "conservation of that race which has given us the true spirit of Americanism is not a matter either of racial pride or of racial prejudice; it is a matter of love of country." Rejecting what he called the "political sophistry that all men are born with equal character to govern themselves," Osborn urged an adherence to the motto "care for the race, even if the individual must suffer."

The partial *T. rex* skeleton on display, and the other magnificent dinosaurs in the American Museum's collection, provided Osborn with an irresistible lure. Visitors interested in the spectacle of prehistoric monsters had to walk through exhibit halls implying that a racial hierarchy not only existed but was as natural and correct as the positions of the bones mounted on each armature. It was a cycle that fed on itself, allowing Osborn's caustic racial influence to grow as dinosaurs expanded into popular culture.

Osborn's insistence on racial hierarchy culminated in his decision in 1921 to offer the American Museum's Hall of Man as the venue for the Second International Eugenics Congress. Researchers from Europe and North America gathered to discuss how to maintain virtuous genetic traits in a world in which global cultures were mingling and mixing. The plight was illustrated by two models. One, built from the average dimensions of one thousand white U.S. Army veterans, was pudgy and short, the implied result of the diminishing returns obtained from mixing different strands of humanity. The other, built from the average dimensions of what was called "the fifty strongest men of Harvard," was tall, muscular and regal, an idealized human form built from conscious breeding of the upper class. The event brought together well-known scientists from around the world, including Alexander Graham Bell, the inventor of the telephone, and Leonard Darwin, the fourth son of Charles Darwin and a former member of Parliament.

Speakers stood at a dais in front of a diorama depicting early modern humans known as Cro-Magnon living in what is now France, a setting that Osborn felt highlighted a "greater artistic sense and ability than have been found among many other uncivilized people." When it was his turn to speak, Darwin argued that "rational methods in human affairs"—a euphemism that meant limiting the reproductive abilities of those he considered less desirable—were the only way to avoid the suffering that "animals in the wild have to endure because of that struggle for existence to which they must submit." Through eugenics, he said, "the end of our species may be long postponed and the race be brought to higher levels of racial health, happiness, and effectiveness."

Osborn understood that the rising influence and popularity of the American Museum gave his theories of racial superiority greater scientific weight, even when eugenics was not the topic at hand. He often opened the museum and its collections to reporters and filmmakers, provided that the institution was treated with the proper respect. In 1912, he allowed a newspaper cartoonist named Winsor McCay to use the museum as the set of a silent movie that began with an open-topped car carrying McCay and a group of friends breaking down with a flat tire in front of the museum's Seventy-seventh Street entrance. McCay, who was best known for his surrealist comic strip *Little Nemo in Slumberland*, heads into the museum with his companions to see the dinosaurs.

While there, McCay bets the others that he can make the *Bronto-saurus* move on its own. A large pad appears and he begins drawing a dinosaur that he names Gertie, who comes to life as a cartoon and starts to obey all his commands. At fourteen minutes long, the film was the first to feature an animated dinosaur. It was an immediate hit when McCay took it on a vaudeville tour, where he held up signs with written dialogue and ended the show by appearing to jump on Gertie's back and ride off into the sunset. Two years later, McCay produced a new version of the film that was shown to packed audiences in some of the country's first theaters designed to show feature-length movies.

The merging of dinosaurs into popular culture left Brown in a sort of netherworld, stuck between the present and the deep past. In New York, he found himself fielding questions from movie studios and authors enticed by the burgeoning popularity of dinosaurs, making the museum's early struggles for relevance and funding seem like a dim memory. Every escape from the city on an expedition felt like heading back not only in geological time, but to an earlier era of his own life, when his concerns were narrowed down to finding the next fossil. At the age of forty, he was no longer a young man, yet he had none of the constraints of adulthood. His only immediate family was a daughter he rarely saw who was cared for by her maternal grandparents; his home was as much the road as it was the small apartment he kept in Brooklyn.

He spent the summer of 1913 competing with the Sternbergs in the Red Deer River Valley, once again hiding from the modern world in pursuit of buried monsters. The spell of escape was broken once the Sternbergs began to find impressive specimens on their own. "Sternberg's sons were up to camp in the boat today and report finding an Albertosaurus skull in fair condition," Brown wrote in a report to the museum. "There [are] plenty of exposures here for all of us for this year but I am really provoked that the Ottawa people should follow our footsteps so closely." His

1913 season was a disappointment by his standards, and he began the 1914 season in the Canadian outback alarmed at the continued success of the Sternbergs compared with his own relatively slow start. "Sternberg and his party are just below us [on the river] and have taken out some fossils from our territory but we have at present no serious disputes. They have no regard for [the] ethics of bone digging," he wrote in a report to the museum that summer.

The field was often his fortress, but in the summer of 1914 he could no longer hide. In early July, a letter from Osborn informed him that "the great European war has just burst, with some uncertainty still as to whether Great Britain will be drawn in. We are fortunate to be out of its direct scope, although doubtless it will affect our affairs in various indirect ways. I can only hope it will not involve the finances of the Museum in any way."

Throughout his life, Brown was a master at reinventing himself into new personas that changed with the times, replacing the Kansas farm boy with a cultured European before turning into a flamboyant playboy at home in New York. In the nineteen years since he discovered the American Museum's first dinosaur specimen at Como Bluff, he had almost single-handedly made it the world's premier collection of fossils. Now, as the war began, he searched for a new way to remain relevant. A letter from the head of the Vertebrate Paleontology Department later that summer warned Brown of the changed world he would return to.

"The war is going to be a long and exhausting struggle, and the longer it lasts the harder we shall be hit by it," he wrote.

Chapter Fifteen

THE MONSTER UNVEILED

IN OCTOBER 1915, A BRITISH ARMY REGIMENT OF VOLUN-
teer soldiers launched a major attack on German trenches dug into
the marshy meadows outside of the village of Loos in northern
France. Few of the men rushing toward battle had any preparation
for warfare, leaving them disoriented in the chaos of the Western
Front. Hoping to break through the enemy's lines, forces under
Field Marshal Sir John French fired cylinders of chlorine gas ahead
of their charge, the first time that the British Army engaged in
chemical warfare. "The gas hung in a thick pall over everything,
and it was impossible to see more than ten yards. In vain I looked
for my landmarks in the German line, to guide me to the right
spot, but I could not see through the gas," a second lieutenant
wrote to his fiancée during a lull in the fighting. The British suffered over fifty thousand casualties that month, nearly double that
of the Germans. Among the possessions of German forces camped
in France were copies of a magazine called *Die weissen Blätter*,
which contained a just-published novella titled *The Metamorphosis*
written by an insurance officer in Prague named Franz Kafka.

That same month, near the Pacific, a recently divorced poet
named Sara Bard Field led rallies for women's suffrage in small
towns through California and Nevada each evening as she

made her way east during a cross-country drive in an open-
air Oldsmobile from San Francisco to Washington, DC, where
she planned to deliver a petition to President Woodrow Wilson
demanding a constitutional amendment that would grant women
the right to vote. Despite being accompanied by a driver and a
mechanic, she found her progress stalled by poor road conditions
and repeated attempts at sabotage. She eventually reached the
White House in early January, with more than half a million sig-
natures. A newspaper editor who was surprised that a prominent
suffragist was physically attractive headlined a story about her
arrival as "Mrs. Erghott Is Not Fat Woman," using the married
name she had abandoned. Not long after she arrived in DC, Field
came up with the popular slogan "No votes, no babies!"

Along the eastern seaboard, the Boston Red Sox defeated the
Philadelphia Phillies in five games to win the World Series in the
span of less than a week, an ending so quick that a young Boston pitcher by the name of Babe Ruth made only one appearance
during the entire series, grounding out as a pinch hitter in his team's
sole loss. Some ten thousand fans stood in Times Square watching
an electric scoreboard that the *New York Times* had erected at the
corner of West Forty-fourth Street and Broadway. "At one time
the crowd was so great that it surged across Broadway. The wonderful mechanical device reproduced the plays almost simultane-
ously with their execution at the Phillies' park at Philadelphia . . .
on every side it was conceded to be the last word in automatic
baseball portrayal contrivance," the *Times* boasted.

And at the corner of West Seventy-seventh Street and Central
Park West, in the only room on the fourth floor that could contain its height and weight, the world got its first glimpse of a fully
mounted *Tyrannosaurus rex*. Gone were the pair of long legs and
a pelvis that underwhelmed visitors when Osborn first unveiled
them in a moment of weakness nearly a decade before. Instead, in
the center of the Hall of the Age of Man, there stood a beast eigh-

teen and a half feet tall that looked as if it were capable of stepping off its pedestal at any moment and leaving a trail of destruction across Manhattan.

Until that time, nearly every impressive dinosaur specimen in a museum exhibit anywhere in the world had been a herbivore, giving the impression that dinosaurs were outsized reptilian cows. That fell away in the presence of a massive carnivore. It was as if a sheet had finally been pulled down, revealing a world that had been hidden from view. Predation, protection, competition, decay; each aspect of the great struggle of life came together in the presence of a creature whose obvious power implied a full and vibrant ecosystem. The slow timetable of preparing the fossils and building a mount had left a large gap between what scientists knew about the former world and how they were able to present it to the public. The *T. rex*, whose dominance was so apparent that it needed no scientific explanation to be understood, narrowed it in an instant. Its strength was undeniable, forcing the recognition that dinosaurs were something more than a novelty and instead evidence that a dominant lifeform's stay on Earth was not permanent. Physically, the beasts were monsters; symbolically, they implied that humans, too, would one day be replaced.

In truth, the specimen was not one animal, but several. With the only three known skeletons of *T. rex* in his possession, Osborn directed his staff to select the best parts of each, combining bones to create a display that was both the remnants of once-living creatures and a sculpture built with human hands. Bits of sandstone harder than granite still clung to parts of the skulls and vertebrae, a reminder of the nearly ten years of grueling work that it took to fully free the fossils from stone.

When completed, the specimen stood erect with its backbone almost vertical, like a kangaroo. Its tail dragged far behind, like the train of a wedding dress. The posture, which museum curators later dubbed a "Godzilla pose," made the beast seem more even overwhelming than its massive jaws suggested, a staggering machine of destruction. (The specimen's positioning would eventually be changed in the 1990s to reflect the fact that its backbone likely stretched out horizontally, with its tail gliding behind it in the air.) Its serrated teeth, each the size of a banana, glistened in the lamplight of the museum hall.

Osborn, finally free of the feeling that he was forever stuck in the wake of the Carnegie Museum, could not contain his sense of triumph. "This skeleton is the finest single exhibit in the department; its mounting technique is considered exceptionally good,—and of its kind unequaled; and the scientific value and popular interest are enhanced by the extreme rarity of these skeletons, their gigantic size and the fierce and predatory character of the animal," he boasted in that year's annual report.

Thousands of visitors stood outside the museum each day waiting their turn to view the beast, undeterred by neither the chilly weather nor the long lines. When it was their turn to marvel at the completed mount, many invariably stepped back in fright, utterly unprepared for the shock of staring up at a towering carnivore. Newspapers across the country covered the unveiling in breathless tones, as if unable to look away from the reappearance of a long-dead monster. Most articles focused on the grim specifics of the beast: the length of its teeth, the weight of its jaws, the ferocity of its appetite. The *New York Herald* devoted a full page of photographs to the new exhibit. "Behold the tyrano as he must have looked in life—except, of course, for the usual upholstering of flesh and hide," it noted. "Books tell us that the carnivorous dinosaurus tyrranosaurus [sic] was a flesh-eating reptile with the tendencies of a tyrant. You better believe it."

In a nearly full-page poem published in the *New York Times*, a humorist by the name of Captain George Steunenberg described his first encounter with the *T. rex* display.

Great shades of Father Adam, and Noah and the Ark!
What's this ungodly thing I see right here near Central Park?
A skeleton of something rearin' way up in the air
Its head cocked catawampus like a bronco on a tear

The poem ended with the refrain:

But then I guess you've had your day and so we'll leave you here
Reflectin' on the good old times of your remote career
May pleasant recollection dwell in that big bony head—
So long, you old hell-raiser: I'm most glad you're dead!

The bones themselves were just part of what made the exhibit so captivating. Next to the specimen itself hung an oil painting by a museum staffer named Charles R. Knight that would become one of the most influential artworks in the museum's history. Knight's unusual path began in Brooklyn, where he was born with severe nearsightedness. A later injury to his right eye in childhood left him legally blind. His compromised vision did little to dent his fascination with animals, however, and at the age of twelve he enrolled in the Metropolitan Art School, which was housed in the basement of the Metropolitan Museum of Art. Eventually staff at the American Museum noticed a thin young man with oversized glasses who could often be found in front of taxidermied animals in the exhibition halls, holding a sketchpad so close to his face that his nose pressed against the paper. The drawings he produced had a sense of buoyancy, as if his pencil was capable of reanimating life. Informed of his talents, Osborn commissioned him to paint prehistoric mammals. Pleased with the work, he brought him on staff and asked him to paint murals depicting dinosaurs in their natural habitat.

Prior to Knight, most illustrations of dinosaurs had shown them as dull, slow creatures, following Sir Richard Owen's conception of the beasts as dimwitted and poorly matched with the world around them. Though he never went on a fossil expedition and had not trained as a scientist, Knight revolutionized the public conception of dinosaurs through beautiful paintings that portrayed alert, vibrant animals in moments of high danger: creeping up on an enemy, leaping with their claws out to slash an opponent in battle or devouring the meat of their prey. "I felt that I had stepped back into an ancient world—filled with all sorts of bizarre and curious things, and in imagination I could picture quite distinctly just what these mighty beasts looked like as they walked or swam in search of food," Knight later wrote in an unfinished autobiography. In time, Knight would be recognized as the pioneer of what is now known as paleoart, a genre that incorporates scientific evidence to imagine distinct moments of drama in the lives of individual dinosaurs, depicting them as present in their prehistoric environment as a wild animal would be in a forest today.

For the 1915 exhibit, Knight painted a lone adult *T. rex* scanning the distance for potential prey. Its back is nearly vertical, making it appear taller than some of the nearby trees. Its eyes seem alive, its body alert. For the first time, the fearful symmetry of the creature's bones disappeared behind a colorful, fully-realized animal, like finally seeing a completed building after looking only at its blueprints. Through art, Knight accomplished what paleontology had only before hinted at, showing the truth of a world populated with animals that were once as real as those at the nearby Central Park Zoo. Osborn had long wanted to dazzle visitors with two *T. rex* skeletons engaged in battle. With Knight's mural, he accomplished something more: viewers could not only see the bones of the most ferocious animal that ever lived, but could walk away with the feeling that they had been transported back in time. There was no trickery or hokum involved, only the ability of a masterful painter to depict a world so complete that modern life shrank away to reveal the churn of billions of years of constant change on Earth.

The combination of art and anatomy prompted a public reaction unlike anything that had accompanied the unveiling of gargantuan specimens such as a *Brontosaurus* or *Diplodocus*. Instead of focusing on size alone, commentators were suddenly struck by the question of what the discovery of the *T. rex* meant for the importance of humankind in comparison. "We should worry. The animal kingdom may be losing in mere bulk and ponderosity," the *Los Angeles Times* noted in an editorial that ran just a few days after the exhibit opened, and searched for a justification for the diminutive features of modern life that were far less physically impressive than the prehistoric past. Clearly, the *T. rex* would win a contest of sheer brawn; mammals, however, were blessed with more brains and beauty, the paper finally reasoned. "Bigness doesn't count for everything, and we have implicit faith in the law of the survival of the fittest. Animal life may be growing 'small by degrees' but it also becomes 'beautifully less.' The tyrannosaurus was as ugly as the Maltese kitten is graceful."

With its hulking power, the *T. rex* seemed to provide the perfect Darwinian fable. The American Museum was besieged with letters and questions, asking for an explanation of how such a ferocious creature with seemingly no natural enemies could have gone extinct. In the search for answers, scientists often projected their own prejudices. "Armored or unarmored, predatory or herbivorous, [dinosaurs] all disappeared, and the mammals entered into their heritage of the earth. Why they disappeared I do not certainly know—probably bigness had something to do with it; probably lack of brains had more," W. D. Matthew, who succeeded Osborn as head of the museum's Department of Vertebrate Paleontology, wrote in the *New York Times*, in an argument that carried a twinge of the eugenic movement's obsession with intellectual ability. Unlike other large mounted dinosaur specimens, the *T. rex* seemed to function as a celebration of human intelligence, the rare scary thing that made a person feel better about themselves. Had it been alive, it could easily devour a person with its teeth larger than a human brain. Yet it was extinct and humans were not, a scoreboard that seemed to confirm the superiority of reason and intelligence above raw power.

In time, evidence that a meteor crashed into the Earth and rapidly changed the climate would undercut the notion that brains had anything to do with the fate of *T. rex* and other dinosaurs. (Indeed, contemporary scientists now believe that the species was roughly as intelligent as the chimpanzee.) But at that moment, the beast provided the symbol that Osborn had long been looking for to support his conceit that eugenics, with its misbegotten notion that intelligence was limited to heredity, could help humankind avoid the same fate. He had his monster, and the only thing left to do was stand and watch the crowds line up to gawk at it, his noxious influence growing more powerful by the day.

THOUGH THE WORLD MIGHT NEVER have known of *T. rex* if not for his willingness to brave the canyons of Hell Creek, Barnum Brown did not linger in New York after its unveiling. Within a few months he was back in northern Montana, trying to pinpoint the exact layers of sediment above which no dinosaur fossils could be found. He often retreated into the badlands as a way of keeping the modern world and its responsibilities at a distance. Still, he could not deny the reality that his life was changing.

Before he discovered *T. rex*, he was a scrappy young man from Kansas whose only dream was to find enough fossils to allow him to continue a life of adventure. Since then, he had scaled the heights of his profession, discovering the vast majority of the bones on display in an institution now known around the world for its collection of dinosaurs. Yet he had also suffered unspeakable loss, which he was reminded of every time he slept with only the memory of Marion as comfort. Time seemed to be running away from him, the present more and more foreign. In a letter to Osborn, he confessed that a conversation with a local rancher made him feel like Rip Van Winkle, woefully out of date with the modern age. "People tell me of a 'man who years ago took out big mastodons in the breaks'," he wrote, both amused and saddened to realize that the man they were referring to was his younger self.

No matter how much he tried, he could never quite escape the present. Every so often, he traveled up to Oxford, where he would stay "for a few hours to see his daughter and settle finances with his father-in-law," Frances later wrote. "She was aware of what her father did the days and months when she did not see him. In fact, she had her own name for his occupation, dubbing him a 'dig-boner'." After a visit he would often flee to the field, as if trying to keep the sense of responsibility off his scent.

That avenue of avoidance would soon close. The American Museum canceled its three planned field expeditions the following year, keeping Brown out of the dinosaur beds for the first time since Marion was pregnant. "The uncertainty as to how the entry of the United States into the world war might affect the affairs and staff of the Museum made it advisable to postpone fieldwork," Matthew wrote in the annual report in 1917. Instead, Brown did what came naturally to a man for whom staying in one place felt like hell: he became a spy. In notes for his never-published biography, Brown wrote that he accepted a wartime role for the U.S. Treasury Department which he would only describe as "establishing depreciation and depletion of oil properties for taxation purposes." Charting the world's oil reserves relied on the same set of skills that allowed him to read rock formations. With a glance he could tell where sediment layers ended and identify the locations that were the most likely to contain oil fields below. Hunting natural resources and mineral wealth had once pushed miners in Europe deep into quarries from which they reemerged with strange bones; now, decades later, Brown reversed that process and used his abilities honed finding fossils to help the U.S. government locate and secure energy reserves for the war. In time, Brown would accept missions that took him to Turkey and other locations around the world, providing Washington with maps of likely untapped oil fields.

His clandestine work continued after the museum resumed funding digs, and for the remainder of his career he maintained an association with the many tentacles of the federal government. Sometimes the work was for private companies, and he shared the information gathered with the government; other times the work was explicitly for the government, meant for no one else's eyes. Often he found himself working on expeditions that served both masters, such as a trip to Cuba in 1918 during which he uncovered fossilized mammals from a hot spring and evaluated potential mining sites for copper.

He continued to push himself past men half his age, while the friends and rivals he had worked with during the early stages of his career retired or settled into desk jobs at prominent museums. "I saw Dr. Barbour not long ago . . . he thinks you are working too hard," Matthew wrote to Brown that spring, after learning that he was spending three hours a day building a machine to drain water out of a hot spring. "The point is this—your health . . . must override any other consideration. As a mere matter of policy it would be the worst sort of mistake to risk damage . . . as you are past the age of easy recuperation from damage."

✢ ✢ ✢ ✢

WHILE BROWN CONTINUED TO PURSUE adventures that took him as far as India and Burma in search of fossils, his most famous discovery took on a life of its own. In 1914, a former newspaper cartoonist turned marble cutter named Willis O'Brien built a miniature boxer out of clay as a gag to break up the monotony of an afternoon. Another worker in the shop noticed what he was doing and built his own boxer, and the two staged a mock fight. O'Brien began to wonder if he could make it appear as if sculptures were moving on film if he stopped and repositioned them one frame at a time. With the help of a newsreel photographer, he climbed to the roof of the Bank of Italy building in San Francisco and filmed a crudely-made dinosaur and caveman jousting for survival. Looking to put this newfound visual sleight of hand to work, he made a series of five-minute films with titles like "Rural Delivery, Million B.C." for the Edison Company, his skills improving with each completed product.

In 1918, O'Brien wrote to Barnum Brown asking for technical help on a movie he planned to direct called *The Ghost of Slumber Mountain*. Brown, who was increasingly aware that his hardscrabble life of prospecting now offered fiscal rewards beyond the frontier, readily agreed, and tutored O'Brien in subjects ranging from how dinosaurs likely moved their bodies to how they traveled in packs. With his input, O'Brien captured the most realistic dinosaurs yet displayed on film. In the story, a man named Uncle Jack tells his nephews a tale of seeing prehistoric creatures through a magical telescope carried by a man named Mad Dick. When *The Ghost of Slumber Mountain* came out, its special effects outshone all of its human stars (including O'Brien, who decided to play Mad Dick himself). Advertisements plastered across the country tempted audiences to come see as "These giant monsters of the past are seen to breathe, to live again, to move and battle as they did at the dawn of life!" The film's climax featured a battle between a *T. rex* and a *Triceratops*, the first time that a *T. rex* was a screen villain. The film brought in slightly more than $1.8 million in today's dollars, a profit more than thirty-three times the cost of its production.

That taste of success ramped up O'Brien's ambitions. Over the next seven years, he toiled away on an adaptation of Conan Doyle's novel *The Lost World*. It was the first full-length film to pair human actors with stop-motion dinosaurs, fulfilling every child's fantasy. In a publicity stunt, Sir Arthur Conan Doyle screened clips from the not-yet-completed film to an audience that included Harry Houdini at a meeting of the Society of American Magicians in New York. Conan Doyle, whose faith in the supernatural led him to consult fortune-tellers and believe that fairies existed, initially refused to say whether the dinosaurs in the film were simply special effects. In private, he admitted to Houdini that "In presenting my moving dinosaurs I had to walk warily in my speech, so as to preserve the glamour and yet say nothing which I could not justify as literally true. . . . I could not resist the temptation to surprise your associates and guests. I am sure you will forgive me if for a few short hours I had them guessing."

When the film premiered in 1925, it featured scenes so lifelike that some audience members believed the dinosaurs were real. Herds stampede; a pterodactyl strips meat from a fresh kill; a bullet wound in the flesh of an allosaur appears to gush blood. O'Brien again featured a *T. rex* as the ultimate monster, replacing the allosaur in Conan Doyle's original text. For a model, he drew from Charles Knight's painting in the American Museum. Where other dinosaurs grapple and battle, the film's *T. rex* is so fierce that it first rips off the leg of its prey and then begins eating it alive. The film's popularity made it the archetype for monster movies, with the *T. rex* as the star. In a review, the *New York Times* noted that "some of the scenes are as awesome as anything that has ever been shown in shadow form." *The Lost World* became the first commercial film to be shown as in-flight entertainment on an airplane, distracting passengers on a flight that left London's Croydon Aerodrome in April of that year.

Among those who watched the film was Merian C. Cooper, a former Air Force pilot who escaped from a prisoner-of-war camp and returned to civilian life with a plan to join the nascent motion picture business. While working at RKO Studios in Hollywood,

he came across a soundstage featuring O'Brien's models of dinosaurs and decided to jettison a project he was working on about
baboons and instead focus on an idea in which a giant ape fights a
Tyrannosaurus. The completed film, by then known as *King Kong*,
appeared in 1933 and proved a huge success, pulling in more than
$1.5 million in today's dollars during its first weekend despite the
ongoing Great Depression. More than six thousand movie-goers
attended each sold-out showing at the newly-opened Radio City
Music Hall, where audience members could be heard yelling and
whistling when Kong faces down the *T. rex* before being kidnapped and brought to New York. "Human beings seem so small
that one is reminded of Defoe's 'Gulliver's Travels'," noted the *New
York Times* in a review of the film.

Brown made his first trip to Hollywood a few years later after
receiving a call from Walt Disney, who planned to follow the
commercial success of Mickey Mouse and Snow White with a film
that he expected to "change the history of motion pictures." He
asked Brown to educate animators at his studio about everything
associated with dinosaurs, from different geological formations to
the relationships between various species to the most up-to-date
theories on why and how they became extinct. When Brown traveled out to Hollywood to give his thoughts on models and scenes
sketched out for the upcoming film, he "found their walls covered
with very credible and accurate types of prehistoric life including
the associated floras," he later said in a lecture to the New York
Academy of Sciences.

The finished product appeared in November 1940 as a segment
of Disney's masterpiece, *Fantasia*. The film's narrator explains that
audiences will soon see "a coldly accurate reproduction of what
science thinks went on during the first few billion years of this
planet's existence." As an orchestra plays Igor Stravinsky's *Rite of
Spring*, microscopic blobs split, sea creatures form and dinosaurs
appear on a rugged, unsettled planet. Suddenly, rain begins to fall

and a herd of herbivores look up to see a *T. rex* coming for them as lightning crashes behind it. It catches up to a *Stegosaurus*, which fights it off with its spiked tail until the *T. rex* bites its neck, killing it. The other dinosaurs leave as the *T. rex* begins to devour its prey. Soon, however, the swampy world of the *T. rex* is replaced by a parched land full of downed trees and mud, and then only footprints leading to fossils. The segment offered millions of children their first impression of dinosaurs, centering the action on the power and relentlessness of the animal that still awed visitors daily at the American Museum.

Unlike anything else found in a natural history museum, the *T. rex* morphed into a staple of popular culture. In the pages of comic books Superman fought one, Wonder Woman rode one and Batman captured a mechanical one and kept it as a souvenir in the Batcave. At the Chicago World's Fair, an animatronic *T. rex* stood in the center of an exhibit sponsored by the Sinclair Oil Company, its gleaming rows of teeth terrifying young and old alike. In whatever form it took, the instinctive fear that the creature conjured—a feat unmatched by specimens of larger herbivores—made it a magnet, holding an audience's attention long enough to build a popular acceptance of concepts ranging from geology to the first mainstream depictions of climate change.

Brown would often remark that nothing else he had done in life came anywhere close to its importance. His association with the species led to appearances on radio shows and television, turning him into one of the first celebrity scientists. Each week, millions heard lectures he delivered on CBS Radio, while millions more each year viewed the specimens he had uncovered, a combination which seemed to fulfill a prophecy of showmanship handed down from his namesake, P. T. Barnum.

For all of the attention the *T. rex* brought him, there was one audience whom Brown continually shut out. As she grew into a young woman, Frances knew her father only through his absence, ed that his silence signaled that he was still alive. Her faith

rewarded. After receiving no communications from him

h during a mission searching for oil in what is now Ethi-

Anglo-American Oil Company presumed he was dead.

oil company had given up hope for its explorers' sur-

useum felt that it must pass this sad information along.

Barnum's young daughter did not believe a word of

later wrote. During the trip, a French woman with

n may have been having an affair learned of Frances's

d gave him a medieval Coptic cross made of brass to

ughter. Brown held on to it for many years, passing it

only after she became an adult.

Frances often wrote to the American Museum seek-

ion about her father. Once, she asked for a private

n's discoveries, as if by spending time with them she

rstand what was so compelling that it made him cast

W. D. Matthew, the head of the Department of Verte-

ntology, replied that he would take "great pleasure [in

the daughter of my old friend and associate and I shall

ate to anyone else the privilege of showing her around a

tthew wrote a short note to Brown in the field whenever

visited the museum, each time forcing the present into the

ast.

ould take the start of another world war—and Brown's fears

he *T. rex* might be destroyed—to bring the two halves of his

ogether.

Chapter Sixteen

A SECOND
CHANCE

THE PLAY-BY-PLAY ANNOUNCER STAMMERED AS HE READ
from a piece of paper shoved into his hands at the Polo Grounds
shortly after two in the afternoon, interrupting a football game
between the New York Giants and a short-lived team called the
Brooklyn Dodgers. "The Japanese have attacked Pearl Harbor
Hawaii, by air, President Roosevelt just announced," he said, his
voice rebounding throughout the stadium on the blustery afternoon of December 7, 1941. Throughout the announced crowd o
55,501, young men stood up and made their way to the exits, pre
paring to report for duty.

The reality of war fell on the city like a sudden rain. At th
Brooklyn Navy Yard, heavily armed guards set up checkpoin
at the dry docks where two 45,000-ton battleships were und
construction. Police officers surrounded the Japanese consulate o
Fifth Avenue, where they could detect the smell of burning pape
Fighter planes from Long Island's Mitchel Field hummed along th
shoreline, the tinny sound of their propellers echoing through th
canyons of Lower Manhattan.

There was a sense that the enemy could appear at any moment,
recreating the terror of Pearl Harbor along Park Avenue or the
Brooklyn Promenade. New Yorkers should not "feel entirely se

because you happen to be on the Atlantic Coast. There is no comfort in that," Mayor Fiorello LaGuardia warned in a radio broadcast from his desk at City Hall that was aired on five stations. Two weeks later, *Life* magazine published a detailed sketch showing a squadron of Nazi aircraft approaching the city from the southeast, under the headline "How Nazi Planes May Bomb New York."

In those early days of the war, an invasion seemed inevitable. Japanese forces had crossed the Pacific and left the military complex at Pearl Harbor a smoking ruin; what would prevent them from marching farther forward and inflicting the same damage on San Francisco? Germany had hollowed out London after a barrage of fifty-seven consecutive nights of bombing, and it seemed only a matter of time before the city would fall. Nazi submarines operated at will in the Atlantic, sinking any hopes that the width of the ocean could protect the United States from warfare.

Throughout the country, Americans began the glum process of boarding up and hiding the objects that mattered most. In Washington, DC, a congressman from Michigan named Fred Bradley proposed that all of the gleaming white marble buildings in the nation's capital be repainted dark gray to make it more difficult for enemy aircraft to see them. Workers at the Museum of Modern Art in New York began pulling down paintings from the third-floor galleries and putting them in a sandbagged storeroom each night before rehanging them each morning. On the Upper East Side, the Frick Collection painted its skylights black. Museum workers across the city received memos sent out by building engineers on how to respond in the event of a bombing, either by air or by an explosive hidden in a bag. In the case of an attack, workers should immediately head to the exhibition floors and "gather up shattered fragments [of an artwork] and wrap in cloth marked with collection number." To ready itself for a sustained bombing of the city, the American Museum of Natural History developed plans to turn its complex of buildings into a vast public shelter, going so far as to make a deal to purchase pianos in bulk to entertain and calm those it expected to host huddled inside.

Priceless pieces of art and culture were nailed into crates and rushed away from the coasts. Seventy-five of the most important paintings and sculptures in the collection of the National Gallery of Art—including three Raphaels, three Rembrandts and Gilbert Stuart's portrait of George Washington—were loaded onto a train and sent to storage at the sprawling Biltmore estate, the largest private home in the country, tucked in the mountains near Asheville, North Carolina. Some fifteen thousand items from the Metropolitan Museum of Art, filling up ninety truckloads, were hidden at an empty mansion outside Philadelphia. Museum walls in San Francisco and San Diego turned bare as collections were packed and taken to protected vaults in Colorado Springs.

At the American Museum of Natural History, the size and weight of most objects on display made the question of hiding them impossible. One dinosaur specimen alone could fill up more than a dozen truckloads, making a wholesale evacuation of the collection the size and scope of a military operation. The museum did what it could at the edges, packing up precious materials such as gold and diamonds that were easy to transport. The dinosaur collection was left untouched, a silent prayer that a lone German bomb would not destroy what had persevered for millions of years. Every specimen remained in place, ready to brave whatever the war brought.

All except for one. Since Brown found the first *T. rex* nearly thirty-five years earlier, no one else had duplicated his feat. The three *T. rex* specimens in the American Museum remained the only relics of the creature known to science. Should the museum suffer a hit during a raid in New York, all evidence of what was now the world's most recognizable dinosaur could be lost. "We had hoped that at least one specimen would be preserved," Brown, who was curator of the Vertebrate Paleontology Department at the time, later wrote. Fifteen boxes of bones containing the first *T. rex* uncovered in Hell Creek soon arrived at the Carnegie Museum in Pittsburgh, which had agreed to purchase the specimen for roughly $100,000 in today's dollars.

The sale of the priceless *T. rex* holotype—the term given to the first example of any species found—to a rival institution would have been inconceivable a generation earlier. Osborn, however, had died in 1935—living just long enough to praise the rise of Nazism in Europe and Hitler's attempts at putting the racist eugenics principles he endorsed into practice. He retired on January 1, 1933, as the most powerful person the American Museum had ever known, twenty-five years after assuming its presidency. In that time, he sent expeditions to every continent on the globe, authored more than one hundred scientific papers, planned a grand new museum entrance on Central Park West in honor of his childhood friend Theodore Roosevelt and increased the museum's endowment by nearly $200 million in today's dollars. Yet it was in ways unlikely to be noticed by a visitor that his influence was most greatly felt. For a generation, every display had to meet his approval, allowing his opinion that racial hierarchy was a scientific fact to infect exhibits ranging from a diorama depicting a seventeenth-century meeting between the Lenape and Dutch settlers in what was then known as New Amsterdam to the presentation of the *T. rex*.

The final years of Osborn's life were consumed with his search for the origin of what he called the white race and his fear that it would be extinguished. He railed against birth control in favor of what he called birth-selection, arguing that by taking steps to engineer the next generation the country could permanently solve its problems of unemployment and overpopulation. Birth control alone "is fraught with danger to society at large and threatens rather than insures the upward ascent and evolution of the human race," he wrote in the *New York Times*, consumed by the notion that the number of desirable white babies would fall behind those of less favorable stock. "Not more but better Americans," he wrote.

While most biologists and anthropologists agreed that the first humans likely lived in Africa, Osborn continued to argue that white Nordic Protestants could trace their lineage to Asia, where, he claimed, the fossil record indicated the presence of humans living 1.5 million years ago with the same brain capacity "equal to that of at least three races living today." At a time when science was often used to backstop racism—no less than the *New York Times* published a three-part essay edited by Osborn under the headline "Whence Came the White Race?" arguing that a "superior breed" whose lineage traced down to American colonists once conquered prehistoric Europe and drove lesser forms of humans to extinction over the course of a few generations—the intensity of Osborn's vitriol was noted by his peers. By writing a preface to Madison Grant's *The Conquest of a Continent, Or the Expansion of Races in America*, Osborn tied his reputation to "about the most uncompromising and aggressive plea for the maintenance of a Nordic and Protestant America, racially and nationally pure and undefiled, that has ever found its way into print," wrote Dr. William Macdonald, a professor at Yale.

Osborn's death extinguished the final embers of an era in which the American Museum was still new and unproven, struggling to appear relevant in the cacophony of New York. Thanks largely to its collection of dinosaurs, it had become the destination for millions of school-aged children on one of their first field trips, a place where they could experience the vast scope of the natural world in person. Until his death, Osborn hung on to the fact that his institution was the only one in the world that contained *T. rex*, the most famous dinosaur on Earth.

As New York prepared for a possible German aerial attack, Brown was among the last people at the museum who remembered the cowboys and prospectors whose work filled its shelves and the race between museums to find and display the most impressive dinosaurs. When the Carnegie unveiled the specimen it purchased from the American Museum in a "Godzilla pose" in 1942, it was the first time that a *T. rex* stood in any place other than New York City in 66 million years. The local press boasted of the museum's "new baby," calling it "the more spectacular of all the exhibits in the Gallery of Fossil Reptiles." The same year, Brown turned sixty-five, the mandatory age for retirement at the institution where he had worked since he was in college. Out of respect for his contributions he was given the title Curator Emeritus and an office—his last connections to the only job he had ever known.

BROWN WAS NOT BUILT TO slow down. Well into his fifties, he embarked on expeditions that left others struggling to keep up. For company, he often brought along his second wife, a New York socialite and author named Lilian McLaughlin. Brash and accustomed to the spotlight, Lilian was the opposite of his first wife, Marion, in nearly every way. Yet in her love of attention— and apparent willingness to indulge in affairs—she was more like her husband than he perhaps recognized. Together, they trekked through Southeast Asia, their path dictated by the needs of the museum. In Pakistan, Brown attempted to find the remains of an animal now known as *Paraceratherium*, an early rhinoceros that was among the largest land mammals to have ever lived. In central Burma, he disappeared for several days in the jungle after missing a fork in the trail and discovered a glowing spider that darted away when he tried to grab it. "Many nights I searched in the jungle and questioned natives and white officers who had passed through that district, but apparently no one else had reported a luminous spider, nor can I find any record of any known elsewhere," he later wrote.

A bout of malaria that left Brown with a fever of 106.2 degrees was among the few things that could stop him. Lilian packed him in a bathtub full of ice and administered massive doses of quinine as he rambled in delusions. "It was mostly of his youth that he spoke. Sometimes he was a small boy again, wandering over the coal mounds on his father's Kansas farm, collecting his first precious specimens," she wrote in a memoir of their adventures titled *I Married a Dinosaur*. "Finally, the words would turn into meaningless whispers that ended in silence, or suddenly jumble together and be lost in his ravings." After six days, the fever broke. A shell of his former self, Brown weighed less than a hundred pounds and was too weak to walk more than a few paces. Yet within a few weeks, he was strolling through a nearby garden. "Before we knew it, the man was dressing in his new Palm Beach suit and wanting to go places," Lilian wrote.

After each expedition, Brown returned to the sanctum of the American Museum. Not long after they were married, he took Lilian on a private tour of the dinosaur halls, where, "when Barnum explained them, speaking as one would of old friends, [the specimens] seemed to change and warm into life," Lilian wrote. "Strolling further down the hall, neither of us spoke. An ageless silence seemed to enshroud that incredible assemblage once flesh and blood, now stony reminders of Nature's magnificent experiment with bulk and brawn. And suddenly it dawned on me as never before why my husband was so obsessed with his work. It was a great work. *He* had done this. The amassing of these prehistoric wonders had been chiefly his doing, and it was not a small thing."

The drive to do great work did not go away with age. Not long after his retirement in the summer of 1942, Brown received a call from Colonel William J. Donovan. A few months earlier, Donovan had been named chief of the newly-created Office of Strategic Services, a military agency that was the forerunner of the modern Central Intelligence Agency. Donovan was the head of a sprawling network of more than twelve thousand people, working both within the government and outside it to understand the Nazi war machine and ready occupied Europe for the eventual landing of American soldiers. He turned to Brown for assistance in planning a possible invasion route via the Aegean Sea, based on Brown's experience prospecting for fossils in the region. Brown jumped at the prospect of another adventure. He left Lilian at their apartment on Broadway, around the corner from the museum, and rushed to wartime Washington, where he spent his days detailing the geological formations and features of the Greek islands.

There was no time for him to find his own place to live in Washington, so he moved in with the person whom he had been avoiding for much of her life: his daughter, Frances. He unpacked his bags in an empty room in her apartment downtown, the first and only time that the pair had lived together since she was an infant. She had been in Washington for over a year, working in the editorial office of the American Red Cross, after the junior college where she taught English literature for seven years closed due to the war. When they stood next to each other, the resemblance between the two was undeniable. Frances shared her father's soft eyes, round face and lips that turned up at the edges, as if always ready to break out in a laugh. But that was as far as the similarities went. She was in nearly every way his foil—a thoroughness that could have come about only by choice. Where he once jumped aboard a boat to Patagonia with two hours' notice, she directed church choirs; where he delighted in breaking rules, she once worked as a college dean. The gap between their approaches to life was so great that it seemed as if Brown had served as a sort of alternative compass, granting Frances security in feeling that if she remained in opposition to her father's far-flung ways, she would always find shelter.

There was no denying Brown for long, though. Like her mother before her, Frances slowly found herself swept up in her father's energy. One night they would go to a party, the next a concert and then a reception the following evening only to repeat the cycle again, a storm with Brown always at its center. Each day would be filled with work, each night with drinks and a sense of possibility. It was as if Frances had opened her home to a twenty-something, not an aging, weathered man who was nominally retired.

Within a few weeks of his arrival, Brown met "a gorgeous and delightful blonde beauty with a Teutonic accent who had no trouble at all in completely captivating Barnum," Frances wrote in her memoir. Soon, all of her father's time outside the office was spent with the new woman, leaving Frances once again on the sideline of his life. "Not that Frances was bothered by Barnum's affairs with women, but she quickly learned that this particular lady was, in all probability, a very competent Nazi spy," she wrote. She warned her father that his new companion might be attempting to get him to spill secrets that could affect the war. "No amount of reasoning seemed to penetrate, and so Frances spent a good many weeks dreaming of an international incident before the lady, somewhat suddenly, departed from the Washington scene and, presumably, went back to Germany not as successful as she hoped to be," Frances wrote.

Like all of Brown's adventures, his time in Washington was short-lived. He spent less than twelve months in the capital before joining the Board of Economic Warfare in 1943, where he turned his attention to completing an aerial survey of Alberta to locate potential oil fields. Upon his return, he went to work for the military by scanning photographs taken by spy planes over areas where he had once prospected in Africa, India and the Mediterranean islands, searching for signs of enemy camouflage. By that time, Frances had left Washington for a job at another college.

Though brief, the time spent in his daughter's company finally opened Brown up to the possibility of a relationship. He brought her along on expeditions to Guatemala and Montana, answering her unspoken question of what she had been missing while in the care of her grandparents as a young girl. In the final years of his life, Frances became one of Brown's closest confidants, the one person besides Marion who could tame a man who for decades had defined himself by running away in search of something new. In 1962, the Sinclair Refining Company asked Brown, then eighty-nine years old, to supervise the construction of nine life-size fiberglass dinosaurs planned for its Dinoland exhibition at the 1964 World's Fair, where the dinosaurs would stand within sight of models of Saturn V rockets and pavilions housing early computers and modems. Fifty million people were expected to attend. Brown commuted daily by limousine from Manhattan to Hudson, New York, to ensure that the anatomy of each specimen was accurate. It seemed only fitting that a man so connected to the deep past would play a small part in a celebration of the future. "This was sheer joy for him. Who else at eighty-nine years of age could command an important, *new* job," Frances wrote.

As he rode each day along the Hudson River, he often had Frances at his side. Their conversations ranged widely, touching on everything from proposed modifications to the exhibition's plans, to how the models would be loaded on a raft to take them to Queens, to any new idea that he had. But most of those concerns soon drifted away, lost in a spell of unhurried moments together as Frances listened to her father tell stories from his life in the prospecting fields. "She never forgot the innumerable things he told her of his past exploits, nor his gleeful speculation about the probable reactions of onlookers when the dinosaurs were eventually floated down the Hudson River to the fair site," she wrote.

Brown did not live to see his dinosaurs delight the city that had made his dreams possible. Two weeks before his ninetieth birthday, he laid down his fork at the dinner table and told Lilian that he was very tired. He slipped into a coma that night and never recovered, dying in St. Luke's Hospital on February 5, 1963. He was buried in Oxford, New York, next to his first wife, Marion. His death was noted in a short article in the *New York Times*, surreally placed next to an illustration of sketches by the fashion designer Yves Saint Laurent. "Dr. Brown was a tireless fossil hunter and was known as the Father of the Dinosaurs because of his successes during the nearly seven decades in which he served the museum," the paper noted, a eulogy that Brown would no doubt have enjoyed given his difficulties in completing a doctorate at Columbia.

During his life, Brown was widely recognized as the best dinosaur collector who ever lived. He went out into the unknown and came back with new puzzle pieces that told the story of life on Earth, and he did it again and again and again, a run of success and discovery as impressive as navigating the stars. At his death, more than half of the dinosaur specimens on exhibit at the American Museum of Natural History were the result of his work, a priceless collection that turned it from an afterthought into one of the most vital museums in the world. "He has discovered many of the most important and most spectacular specimens in the whole history of paleontology," a fellow collector, Roy Chapman Andrew, wrote in a foreword to Lilian Brown's first book. "When he ceases to look for bones on this earth, the celestial fossil fields may well prepare for a thorough inspection by his all-seeing eyes. He'll arrive in the Other World with a pick, shellack, and plaster or else he won't go."

The New York World's Fair opened on April 22, 1964, a little more than a year after Brown's death. Visitors who walked the grounds in Queens could catch a glimpse of the first Ford Mustang, ride on a Ferris wheel in the shape of a giant tire or try to get the song from Disney's "It's a Small World" ride out of their heads. The fair was an assault on the senses, the only time in the history of humankind when a person could watch trained dolphins toss plastic oranges into an audience and within minutes board a slow-moving walkway to view Michelangelo's *Pietà* positioned behind bulletproof glass. For a brief moment, it was as if P. T. Barnum's dime museum had been reborn for the modern age. And in the middle of the fair, staring directly at the giant Ferris wheel, stood a 20-foot-tall celebrity. Fifty-nine years after Barnum Brown first uncovered it in the Montana badlands, a modern monster reigned in the shadow of the Manhattan skyline: *Tyrannosaurus rex*.

THE MONSTER'S TRACKS

THE SIDEWALKS WERE UNUSUALLY EMPTY FOR A MID-September afternoon in Rockefeller Center. Gone were the office workers who would normally be found seeking out the last vestiges of the summer sun, like squirrels trying to save acorns for the winter. No packs of tourists congregated outside the studio windows of the *Today* show. No one was lined up for a tour of 30 Rockefeller Center, or for a show at Radio City Music Hall. For six months, life throughout the country had largely retreated indoors in response to the coronavirus pandemic, leaving the city streets quiet. And yet for those who found themselves walking outdoors at the corner of Forty-ninth Street and Fifth Avenue in the late summer of 2020, there stood a 13-foot-tall treat: a fully-mounted *T. rex*, one of the most complete specimens ever found.

The creature was frozen mid-roar in the windows of the auction house Christie's, which redesigned its public galleries so that the *T. rex* could be seen from outside. Known as Stan in honor of the amateur paleontologist named Stan Sacrison who discovered it in the Black Hills of South Dakota in 1987, the approximately 67-million-year-old specimen was unique not only for the relatively pristine condition of its bones, but for what they suggested about its life. Punctures in its skull and fused neck vertebrae hinted that it had survived an attack from another *T. rex*, perhaps the only animal capable of inflicting such damage on an eight-ton creature of destruction. Christie's estimated that the bones would go for up to $8 million, an unheard-of amount for a natural history object that put it out of the reach of most museums in the country.

The *T. rex* fossil was deliberately priced like a piece of art, not a specimen of science. "There simply aren't *T. rexes* like this coming to market. It's an incredibly rare event when a great one is found," said James Hyslop, head of Christie's Science and Natural History Department, before the sale. "*T. rex* is a brand name in a way that no other dinosaur is. It sits very naturally against a Picasso, a Jeff Koons or an Andy Warhol." In the October 6, 2020, auction catalog for what Christie's called the Twentieth Century Evening Sale, photos of the *T. rex* were found just a few pages away from a Jasper Johns with a high estimate of $1.8 million, a Picasso from a private European collection with a high estimate of $30 million and a Mark Rothko with a high estimate of $50 million.

The idea that science and culture were natural rivals was made plain with the establishment of the American Museum of Natural History and the Metropolitan Museum of Art on opposite sides of Central Park. Yet in time that divide dwindled, like a forest slowly encroaching on an open field. That fossils could indeed be valued like pieces of art was first proven on October 4, 1997. That day, nine people walked into in an auction room at Sotheby's in Manhattan and took their seats. In front of them, the 600-pound skull of a *Tyrannosaurus rex* rested on a cushion. It was but one piece of a specimen, nicknamed Sue, that was at the time the largest and most complete *T. rex* ever found. In the crowd sat representatives from the Smithsonian, the Field Museum and the North Carolina Museum of Natural Sciences, all of whom coveted the prestige and popularity that only a massive *T. rex* could offer.

That a museum would end up with the specimen was not guaranteed. A real estate baron named Jay Kislak was also in the room and prepared to bid, hoping to add Sue to a private collection that already included a 1516 original copy of the *Carta Marina Navigatoria*, the first printed navigational map of the world, and a 1486 edition of Ptolemy's *Cosmographia*, which revolutionized medieval mapmaking by basing its representations of countries and geographical features on mathematical proportions and not their social importance. Sotheby's hoped to steer the specimen to a public institution by giving the winning bidder three years to complete payment, allowing a chance for a fundraising push. "We weren't selling something that should go to a decorator," an executive at the company later said. Yet there were no guarantees. As the auction began, it was the first time that so large a dinosaur would be auctioned off publicly, and the world waited to find out just how much a complete *T. rex* was worth.

Sotheby's had little experience estimating the value of fossils, much less those of a nearly perfectly preserved monster. The auction house officially said that the fossil was worth "$1 million plus," though it had nothing to compare it to, no equivalent Michelangelos or Leonardos to set a floor. At the time, no fossil had ever sold for more than $600,000. In private, auctioneers guessed that it would go for up to $5 million, a figure that some argued was too low and others said was absurd for a single specimen.

"I begin with a bid of $500,000," David Redden, the Sotheby's auctioneer, said in an English accent dulled by decades spent in America. Trim and well-dressed, Redden was used to high-stakes affairs. He had joined the auction house in 1974, and in that time he had stood in the same spot selling all kinds of items, from an original copy of the Declaration of Independence for $2.2 million to several lots of Jacqueline Kennedy Onassis's belongings. But he had never sold a dinosaur.

Of the roughly fifty *T. rex* specimens that have been uncovered since Brown's first discovery in 1905, only a quarter contain more than half of the dinosaur's complete set of bones. Sue, by comparison, was more than 90 percent complete and stretched 42 feet from snout to tail, with an oversized rib cage, as if the creature had just filled its lungs with a deep gulp of air. The specimen was named after Sue Hendrickson, who discovered it while working as an intern at the Black Hills Institute, a commercial fossil dealer. In August 1990, a truck carrying prospectors from the private company broke down with a flat tire while scouting the property of Maurice Williams, a Sioux Indian who lived on a ranch north of Faith, South Dakota. Hendrickson stayed behind with the vehicle while the others left to get help. To pass the time, she went on a stroll, and soon looked up and spotted a gigantic femur and three vertebrae sticking out of a cliff face seven feet above her head. She immediately recognized them as the bones of "a carnivore and definitely big, which for that area could only mean one thing," she later said. The Black Hills Institute paid Williams $5,000 for the right to excavate, remove and retain ownership of the bones.

As preparators at the company studied the bones, they noticed that the creature had sustained a number of painful wounds, a testament to its likely old age. Gashes from serrated dinosaur teeth marked the skull, while several arm bones on its right side were broken. Uric acid crystals were present in the joints, a sign that it likely suffered from gout. Here was not only a massive dinosaur, but one that carried with it the scars of prehistoric life. When the creature died, it had been quickly buried in water and sediment, preserving in its matrix of rock not only its skeleton but the bones of a prehistoric turtle that lived at the same time.

Two years after its discovery, an assistant United States attorney by the name of Kevin Schieffer charged that the Black Hills Institute had stolen the fossil by not obtaining necessary permissions from federal agencies before excavating on land within an Indian reservation. FBI agents seized the plaster jackets containing Sue's bones and locked them under seal in a furnace room at the South Dakota School of Mines and Technology. In 1994, Williams was named the rightful owner of the specimen, and Sotheby's was chosen to sell it on his behalf.

The price surpassed $1 million a few seconds after bidding opened. When it topped $1.7 million, a South Dakota businessman named Stan Adelstein dropped out, noting with pain that his target price had already been exceeded by more than half a million dollars. The bidding continued in $100,000 increments. Among those left were the Smithsonian, whose lack of a *T. rex* was, its curators felt, the only thing keeping the institution out of the elite company of the world's best collections of dinosaurs, and the North Carolina Museum of Natural Sciences, which was the largest natural history museum in the American South and had built a massive war chest to bid on the *T. rex* in the hope that the acquisition would take it out of the league of regional museums and into the company of world-class institutions like the American Museum of Natural History. In addition to the representatives in the room itself, there were also eleven mystery bidders following the auction by phone, adding to the uncertainty of the outcome. "There were rumors flying that Michael Jackson was bidding," the director of the North Carolina Museum later said.

Finally, the North Carolina Museum was the only museum left bidding on the floor itself. Its final offer was $7.2 million; Kislak, the Florida collector, bumped it out of the running with a bid of $7.3 million. Only then did the Field Museum in Chicago, the institution that had once seemed poised to become the nation's premier home of dinosaurs before it abruptly canceled its funding for prospecting during a low point in Barnum Brown's career, come in with an offer: $7.4 million.

"We did not have an iconic specimen, and we wanted one," the Field's president and chief executive at the time later said. Before the auction, the museum had reached out to Fred Turner, who was then the chief executive of McDonald's Corporation—which had its corporate headquarters in suburban Chicago—in hopes that the company would help it acquire the specimen. "I got about 30 seconds into my pitch and Fred said, "You mean to tell me this is the world's most complete, largest *T. rex*? I told him yes. He said, 'We're in,'" the Field's president would later say. McDonald's then asked the Walt Disney Company to join in the bidding consortium. Disney planned to create a life-size cast of the specimen for its Animal Kingdom theme park in Orlando, while McDonald's was preparing to make two casts as part of a traveling exhibit at regional museums around the world.

Kislak raised the stakes once more, bidding $7.5 million. Though that was the Field's predetermined limit, the museum put in one final bid, for $7.6 million. Redden called out for a better offer three times before dropping the hammer, completing the most expensive auction for a dinosaur fossil in history. At a total of $8.36 million after the auction house's fees, a *Tyrannosaurus rex* was now worth as much as works by European masters, spurring rural landowners to try to strike it rich with bone hunters. "After Sue, it was truly frightening," one National Park Service geologist said. "You heard people saying, 'I'm going to give up my blue-collar job and move out West to find my million-dollar dinosaur.'"

Sue was unveiled at the north entrance to the Field Museum's central hall on May 17, 2000. The day before, McDonald's began advertising its tie-in with Disney's animated movie *Dinosaur*, which opened in theaters that weekend. Included in McDonald's prize giveaways were a $1 million check, a trip to a Britney Spears concert and countless Happy Meal toys that included dinosaur squirt guns.

In 2020, there were no such pairings of private companies and public institutions. The field was narrowed in part by the bizarre stipulation that the sale did not include full ownership of its intellectual property, which would be necessary to reproduce the specimen for three-dimensional models or related merchandise that could help recoup the costs. "We would never purchase something unless we owned the rights," said Mark Norell, a paleontologist at the American Museum of Natural History. Still, interest kept rising. Hyslop began the bidding at $3 million. Buyers on phones in London and New York topped that in increments of $100,000. After twenty minutes of flurry, it was over: a private buyer working through the London office of Christie's offered $31.8 million, including the auction house's fees. The buyer's identity was not disclosed, though it was widely believed that the purchaser came from an oil-rich state in the Middle East. That the bid came through London "usually means it's Middle Eastern money, and I know for a fact that there was Middle Eastern interest in the fossil," Norell said.

Whether Stan will ever again be seen in public is not known. During the Gilded Age, business titans such as J. P. Morgan and Andrew Carnegie spent millions to fund expeditions that uncovered and displayed dinosaurs in public museums as a testament to their own prosperity and power. More than a hundred years later, that instinct has shifted toward private displays of wealth. The *T. rex* remains one of the most elusive and sought-after fossils by both public and private collectors. In 2007, actors Leonardo DiCaprio and Nicolas Cage fought a bidding war over the 32-inch skull of a *Tyrannosaurus bataar*, a close cousin of the *T. rex*. Cage won with a $276,000 bid, but a few weeks before Christmas in 2015 returned it by order of the Department of Homeland Security, after it was discovered to have been smuggled out of Mongolia. Nathan Myhrvold, a former chief technology officer at Microsoft, installed a life-size cast of a *T. rex* in the living room of his Seattle-area home, to commemorate his funding of digs that helped uncover nine *T. rex* specimens in Montana at a time when only eighteen had previously been known to science.

Stan's auction price of more than $31 million pushes the likely price of the next *T. rex* specimen that goes on sale even higher, another side effect of the widening barbells of wealth disparity

worldwide. "It's an astounding amount of money," Darla Zele-
nitsky, associate professor of dinosaur paleobiology at the University of Calgary, told the *New York Times*. "I think it would
make it tough for museums to buy fossils, especially at a price
peg like that."

Museums that do not have the means to purchase specimens
on their own are finding themselves in the odd position of acting as showrooms for fossil hunters who own a *T. rex* but do not
have connections with potential buyers. Alan Detrich, an amateur
fossil hunter in Kansas, lent the skeleton of a four-year-old *T. rex*
he found near Jordan, Montana, to the University of Kansas Bio-
diversity Institute and Natural History Museum. Two years later,
he put the specimen on eBay with a starting bid of $2.95 million.
"Most likely the only BABY T-Rex in the World!" he boasted.
The museum asked that Detrich remove from the advertisement
any reference to the specimen's display on its floor. He complied,
though he seemed confused as to why someone would fault him
for selling it. "Well, I own this thing. It is mine. I can do whatever
I want," he told a reporter. Besides, he added, "It's very hard to
reach a billionaire. Putting it on eBay is one way to do it."

Overall, nearly 300,000 potential bidders viewed Christie's
online listing for Stan, shattering the auction house's viewership
records. It is yet one more sign that *Tyrannosaurus rex* remains sur-
prisingly alive for a creature that went extinct 66 million years ago.
Researchers studying the animal now know that, far from being
lumbering brutes, *T. rex* were intelligent creatures that lived in
complex social systems. They were blessed with oversized nostrils
and eyes and an outsized brain case that likely gave them a superb
sense of sight and smell, fashioning them into formidable hunters
at a time when the Earth's temperatures were near their highest
in known history. The eldest known *T. rex* specimen, nicknamed
Scotty, most likely weighed 9.8 tons and was probably in its early
thirties when it died in what is now Saskatchewan, Canada. Its

skull was covered with bumps and ridges down the snout, suggesting armored skin. "It has flair to its face," said W. Scott Persons, a
paleontologist from the University of Alberta, who led a study of
the specimen. The animal's long legs, meanwhile, gave it a stride
that conserved energy even as it ran at speeds of 10 miles per hour
or more, allowing it to use up to 35 percent less energy than other
dinosaurs of similar size, according to a 2020 study. Studies now
suggest that 20,000 adult *T. rex* lived in North America at any
given time. Over the 2.4 million years it was in existence, a total of
some 2.5 billion adult *T. rex* walked the Earth. Given that paleontologists now estimate that only one out of every 80 million *T. rex*
that ever lived was fossilized, Brown's discovery of three of them
becomes all the more astounding.

Over the last thirty years, expeditions in Mongolia and China
have found numerous other types of tyrannosaur that in some cases
were as small as a chicken, increasing the total species count three-
fold. The search for additional family members of *T. rex* is at the
center of what is considered a new golden age of paleontology,
when increasingly sophisticated technology and renewed sources
of funding are fueling a deeper understanding of the planet's past
at a time when its climate is once again rapidly changing. "More
is going on now than ever," Philip J. Currie, a paleontologist at
the University of Alberta, told the *New York Times*. "There were
probably only six of us in the world who were paid," he said, to
focus solely on the study of dinosaurs when he began in the 1970s,
adding, "Right now, there's maybe 150," along with a "colossal
increase in the number of scientific papers."

It's fair to wonder how much of the history of life on this planet
would remain unknown today had Barnum Brown never escaped
the life that was set out for him on his family farm. What would
a modern world without a *T. rex* look like? It's easy to list the billions of dollars in box-office receipts that were brought in by the
appearance of a snarling *T. rex*, the countless children's toys and

pajamas with the dinosaur's image, the tourist attractions ranging from roadside fossil digs in Montana to roller coasters named after the planet's most ferocious beast; all of which would be lost. Yet its influence is greater than that. Without the *T. rex*, it's likely that dinosaurs would have remained little more than novelties, never inspiring the public and scientists alike to imagine a complex former world or prompting them to mine the past for clues on how a rapidly-changing climate once upended the dominant species of the prehistoric past. Every child that attends a natural history museum on a field trip has Barnum Brown—and the *T. rex*—to thank. While some paleontologists resent the attention that a *T. rex* draws compared with other species, the animal's celebrity has a way of bringing science into a conversation from which it might have otherwise been excluded. "People who study non-dinosaurs say dinosaurs get all the attention," said Stephen Bursatte, a paleontologist at the University of Edinburgh. "People who study dinosaurs say theropods get all the attention. People who study theropods say, oh, tyrannosaurs get all the attention."

As temperatures continue to rise today, we can look at the fossils of a *T. rex* and see a glimpse of what may become of humankind if we do not take more steps to address the climate crisis. In the end, that could be the most important lesson we can draw from the fearsome reign of the *T. rex*: in the battle for life on Earth, the climate always wins.

The animal at the intersection of popular culture and modern science still stands on the fourth floor of a building in Manhattan on the edge of Central Park. There, with the noise of New York City reduced to little more than a murmur, visitors find themselves gazing across a 66-million-year gap at the greatest predator that nature ever produced—and whose discovery was the greatest legacy of a man who could not be contained.

ACKNOWLEDGMENTS

ESCAPING INTO DINOSAUR DIGS IN MONTANA—OR AT least pretending to—while working on this book was a welcome diversion as the coronavirus raced through New York City and its suburbs, an area that I now call home. Though libraries were closed and trips into the field canceled, a number of people continued to go out of their way to assist me in my attempt to tell the story of Barnum Brown and the monster he found.

The staff at the American Museum of Natural History were unfailingly helpful and courteous as I came back again and again with questions. Chief among them was Susan Bell, who opened the museum's archives for me and spent several long afternoons in a stuffy records room located in the building's attic while I paged through Barnum Brown's papers and mice scurried past our feet. Dr. Mark Norrell, who, with Dr. Lowell Dingus, wrote a fantastic biography of Barnum Brown, extended his help in finding documents for my own work. And Matt Heenan, in the museum's business office, made the process of finding and republishing photographs from the American Museum's archives incredibly easy.

I would also like to thank several scholars whose work was invaluable to this project. Paul Brinkman uncovered a treasure trove of details of what he called the Second Jurassic Dinosaur Rush, while

Lukas Rieppel artfully examined how dinosaur fossils became a status symbol for Gilded Age tycoons. Adrienne Mayor's work into how folklore incorporated fossils, meanwhile, helped provide a grounding for how humans interacted with dinosaurs before the advent of formalized science, and Deborah Cadbury's work *The Dinosaur Hunters* helped flesh out the stories of early fossil hunters in England.

I am exceedingly fortunate to once again work with a team at W. W. Norton that was supportive of this project from the beginning and continued to make it better at every step of the way. My editor, Jill Bialosky, pushed me to go deeper into Brown's character and offered innumerable suggestions along the way, which helped me get over whatever obstacle was in front of me. I am thankful that Allegra Huston, who copyedited this book, caught several typos and mistakes before they could make it into print and has the breadth of knowledge to both correct a reference to a poem by William Blake and ask if we should include a mention of a glam rock band. Steve Attardo and Faceout Studio conceived of a brilliant book jacket design, while Louise Brockett conjured the perfect subtitle. Drew Weitman, who helped the trains run on time, responded to every issue that came up with kindness and skill.

Special thanks also go to Larry Weissman and Sasha Alper, who helped shape this book back when it was just the wisp of an idea and wouldn't rest until we got its wording just right, and Josie Freedman at CAA.

I am also fortunate to benefit from the support and company of Alan Yang, Jennifer Ablan, Lauren Young, Megan Davies, Dan Burns, Helen Coster, Sam Mamudi, Felix Gillette, John and Carol Ordover, Tony and Maryanne Petrizio, Robert, Gina and Gary Scott, Emily Davis and Ryan and Diane Randall.

And finally, infinite love and thanks go to Megan, Henry and Isla Randall, who made every day a good one regardless of how overwhelming the process of writing a book can sometimes feel. I am incredibly lucky to have you in my life.

SELECTED BIBLIOGRAPHY AND SOURCES

Prologue
THE CENTER OF THE WORLD

Preston, Douglas. *Dinosaurs in the Attic.* New York: St. Martin's Griffin, 1993.

Davey, Colin. *The American Museum of Natural History and How It Got That Way.* New York: Empire States Editions, 2019.

Wilford, John Noble. "When Humans Became Human." *New York Times*, February 26, 2002.

Chapter One
A LIFE THAT COULD CONTAIN HIM

Brown, Frances R. *Let's Call Him Barnum.* New York: Vantage Press, 1987.

Dingus, Lowell, and Mark A. Norell. *Barnum Brown: The Man Who Discovered T. Rex.* Berkeley: University of California Press, 2010.

Brown, Barnum. Unpublished notes. American Museum of Natural History Vertebrate Paleontology Archives.

Cadbury, Deborah. *The Dinosaur Hunters.* London: Foulsham, 2000.

Thomson, Keith S. "Marginalia: Vestiges of James Hutton." *American Scientist* 89, no. 3.

Rabbitt, Mary C. "The United States Geological Survey: 1879–1989." *U.S. Geological Survey Circular.* Washington, DC: Government Printing Office, 1984.

Chapter Two
A WORLD PREVIOUS TO OURS

Cordley, Richard. *A History of Lawrence, Kansas: From the First Settlement to the Close of the Rebellion.* Lawrence Journal Press, 1895.

Brown, Barnum. Unpublished notes. American Museum of Natural History Vertebrate Paleontology Archives.

Bailey, E. H. S. "Samuel Wendell Williston: A Kansas Tribute." *Sigma Xi Quarterly* 7, no. 1 (1919).

Cadbury, Deborah. *The Dinosaur Hunters.* London: Foulsham, 2000.

Murray, J. *Report of the Eleventh Meeting of the British Association for the Advancement of Science.* 1842.

Mayor, Adrienne. *The First Fossil Hunters: Paleontology in Greek and Roman Times.* Princeton: Princeton University Press, 2000.

Mayor, Adrienne. *Fossil Legends of the First Americans.* Princeton: Princeton University Press, 2013.

Kolbert, Elizabeth. "The Lost World." *The New Yorker,* December 8, 2013.

Kolbert, Elizabeth. *The Sixth Extinction: An Unnatural History.* New York: Henry Holt and Co., 2014.

Owen, Rev. Richard. *The Life of Richard Owen. Vol. 1. London,* Gregg International, 1894.

Pierce, Patricia. *Jurassic Mary: Mary Anning and the Primeval Monsters.* London: History Press, 2015.

Dean, Dennis. *Gideon Mantell and the Discovery of Dinosaurs.* Cambridge: Cambridge University Press, 1999.

Conniff, Richard. *House of Lost Worlds: Dinosaurs, Dynasties, and the Story of Life on Earth.* New Haven: Yale University Press, 2016.

Wallis, Severn Teackle. *Discourse on the Life and Character of George Peabody.* Baltimore: John Murphy & Co., 1870.

Winchell, N. H., ed. *The American Geologist: A Monthly Journal of Geology and Allied Sciences.* Minneapolis: Geological Publishing Company, 1899.

Dwight, Timothy. *Memories of Yale Life and Men, 1854–1899.* New York: Dodd, Mead, 1903.

Davidson, Jane P. *The Bone Sharp: The Life of Edward Drinker Cope.* Philadelphia: Academy of Natural Sciences Philadelphia, 1997.

Jaffe, Mark. *The Gilded Dinosaur: The Fossil War Between E. D. Cope and O. C. Marsh and the Rise of American Science.* New York: Crown, 2000.

Brinkman, Paul. *The Second Jurassic Dinosaur Rush.* Chicago: University of Chicago Press, 2010.

Chapter Three
SCRAPING THE SURFACE

Brown, Barnum. Unpublished notes. American Museum of Natural History Vertebrate Paleontology Archives.

Dingus, Lowell, and Mark A. Norell. *Barnum Brown: The Man Who Discovered T. Rex.* Berkeley: University of California Press, 2010.

Dodson, Peter. *The Horned Dinosaurs: A Natural History.* Princeton: Princeton University Press, 2017.

Marsh, O. C. "Notice of New American Dinosauria." *American Journal of Science,* 1889.

Brinkman, Paul. *The Second Jurassic Dinosaur Rush.* Chicago: University of Chicago Press, 2010.

Chapter Four
CREATURES EQUALLY COLOSSAL AND EQUALLY STRANGE

Regal, Brian. *Henry Fairfield Osborn: Race and the Search for the Origins of Man.* London: Routledge, 2002.

Rainger, Ronald. *An Agenda for Antiquity: Henry Fairfield Osborn and Vertebrate Paleontology at the American Museum of Natural History, 1890–1935.* Birmingham, AL: University of Alabama Press, 1991.

Dingus, Lowell, and Mark A. Norell. *Barnum Brown: The Man Who Discovered T. Rex.* Berkeley: University of California Press, 2010.

Gregory, William K. *Biographical Memoir of Henry Fairfield Osborn.* 1936.

Scott, William Berryman. *Some Memories of a Paleontologist.* New York: Forgotten Books, 2017 (reprint).

Chamberlain, Joshua Lawrence. *Universities and Their Sons: History, Influence and Characteristics of American Universities, with Biographical Sketches and Portraits of Alumni and Recipients of Honorary Degrees.* Vol. 1. Boston: R. Herndon Co., 1898.

Peck, Robert McCracken. "The Art of Bones." *Natural History.* May 10, 2012.

Smith, Phillip. *The Popular History of England from the Earliest Times to the Year 1848.* London: A. Fullerton & Co., 1883.

de Chadarevian, Doraya, and Nick Hopwood. *Models: The Third Dimension of Science.* Palo Alto: Stanford University Press, 2004.

Twelfth Annual Report of the Board of Commissioners of the Central Park, for the Year Ending December 31, 1868. New York: Evening Post Seam Presses, 1867.

Proceedings of the Academy of Natural Sciences of Philadelphia 21. Philadelphia, 1869.

Proceedings of the American Association for the Advancement of Science 47. Salem, 1898.

Cain, Victoria. "Albert Bickmore." *Harvard Magazine*, September–October, 2008.

Chapter Five
EMPTY ROOMS

"Disastrous Fire: Total Destruction of Barnum's American Museum." *New York Times*, July 14, 1865.

Barnum, P. T. *Struggles and Triumphs: Or, Forty Years' Recollections of P. T. Barnum, Written By Himself.* Hartford: J. B. Burr and Co., 1869.

Sellers, Charles Coleman. "The Peale Portraits of Benjamin Franklin." *Proceedings of the American Philosophical Society* 94, no 3 (1950).

Kohlstedt, Sally Gregory. "Curiosities and Cabinets: Natural History Museums and Education on the Antebellum Campus." *Isis* 79, no. 3 (1988).

Stulman Dennett, Andrea. *Weird and Wonderful: The Dime Museum in America.* New York: New York University Press, 1997.

Ayres, James J. *Gold and Sunshine: Reminiscences of Early California.* Boston: Gorham Press, 1922.

Rieppel, Lukas. *Assembling the Dinosaur: Fossil Hunters, Tycoons, and the Making of a Spectacle.* Cambridge, MA: Harvard University Press, 2019.

Ross, Delaney. "150-year-old Diorama Surprises Scientists With Human Remains." *National Geographic,* January 29, 2017.

Hermann, A. "Modern Laboratory Methods in Vertebrate Paleontology." *Bulletin of the American Museum of Natural History* 26 (1909).

Rainger, Ronald. *An Agenda for Antiquity: Henry Fairfield Osborn and Vertebrate Paleontology at the American Museum of Natural History, 1890–1935.* Birmingham, AL: University of Alabama Press, 1991.

Chapter Six
A REAL ADVENTURE

Dingus, Lowell, and Mark A. Norell. *Barnum Brown: The Man Who Discovered T. Rex.* Berkeley: University of California Press, 2010.

Brown, Barnum. Unpublished notes. American Museum of Natural History Vertebrate Paleontology Archives.

Rainger, Ronald. "Collectors and Entrepreneurs: Hatcher, Wortman, and the Structure of American Vertebrate Paleontology Circa 1900." *Earth Sciences History* 9, no. 1 (1990).

Wortman, J. L. "Restoration of Oxyaena lupina Copa, With Descriptions of Certain New Species of Eocene Creodonts." *Bulletin of the American Museum of Natural History* 12–13 (1900).

Rieppel, Lukas. *Assembling the Dinosaur: Fossil Hunters, Tycoons, and the Making of a Spectacle.* Cambridge, MA: Harvard University Press, 2019.

Thomson, Kenneth. *The Legacy of the Mastodon: The Golden Age of Fossils in America.* New Haven: Yale University Press, 2008.

Chapter Seven
FINDING A PLACE IN THE WORLD

Jackson, Kenneth T., and David S. Dunbar. *Empire City: New York Through the Centuries.* New York: Columbia University Press, 2002.

Duffus, R. L. *Lillian Wald: Neighbor and Crusader.* New York: Macmillan, 1939.

Riis, Jacob. *How the Other Half Lives: Studies Among the Tenements of New York.* New York: Penguin Classics, 1997 (reprint).

Annual Report of the American Museum of Natural History, vols. 29–32.

Dingus, Lowell, and Mark A. Norell. *Barnum Brown: The Man Who Discovered T. Rex.* Berkeley: University of California Press, 2010.

Brown, Frances R. *Let's Call Him Barnum.* New York: Vantage Press, 1987.

Chapter Eight
THE UTTERMOST PART OF THE EARTH

Dana, Richard Henry, Jr. *Two Years Before the Mast.* New York: Signet, 2009 (reprint).

Meacham, Steve. "Forgotten death at sea stoked Darwin's success." *Sydney Morning Herald*, June 27, 2009.

Brown, Janet. *Charles Darwin: The Power of Place.* New York: Knopf, 2003.

Larkum, Anthony W. D., ed. *A Natural Calling: Life, Letters and Diaries of Charles Darwin and William Darwin Fox.* New York: Springer Science and Business Media, 2009.

Cope, E. D. "Review: Ameghino on the Extinct Mammalia of Argentina." *American Naturalist* 25, no. 296 (August 1891).

Keller, Mitch. "The Scandal at the Zoo." *New York Times*, August 6, 2006.

Gabriel, M. S. "Ota Benga Having a Fine Time: A Visitor at the Zoo Finds No Reason for Protests about the Pygmy." *New York Times*, September 13, 1906.

Dingus, Lowell, and Mark A. Norell. *Barnum Brown: The Man Who Discovered T. Rex.* Berkeley: University of California Press, 2010.

Brown, Barnum. Unpublished notes. American Museum of Natural History Vertebrate Paleontology Archives.

Dingus, Lowell. *King of the Dinosaur Hunters: The Life of John Bell Hatcher and the Discoveries that Shaped Paleontology.* New York: Pegasus Books, 2018.

Brown, Frances R. *Let's Call Him Barnum.* New York: Vantage Press, 1987.

Chapter Nine
BIG THINGS

Rieppel, Lukas. *Assembling the Dinosaur: Fossil Hunters, Tycoons, and the Making of a Spectacle.* Cambridge, MA: Harvard University Press, 2019.

Batz, Bob, Jr. "Dippy the Star-Spangled Dinosaur." *Pittsburgh Post–Gazette,* July 2, 1999.

Nasaw, David. *Andrew Carnegie.* New York: Penguin, 2007.

Rea, Tom. *Bone Wars: The Excavation and Celebrity of Andrew Carnegie's Dinosaur.* Pittsburgh: University of Pittsburgh Press, 2004.

Brinkman, Paul. *The Second Jurassic Dinosaur Rush.* Chicago: University of Chicago Press, 2010.

Wister, Owen. *The Virginian: A Horseman of the Plains.* New York: Macmillan, 1904.

Brown, Barnum. Unpublished notes. American Museum of Natural History Vertebrate Paleontology Archives.

Chapter Ten
A VERY COSTLY SEASON

Brown, Barnum. Unpublished notes. American Museum of Natural History Vertebrate Paleontology Archives.

Dingus, Lowell, and Mark A. Norell. *Barnum Brown: The Man Who Discovered T. Rex.* Berkeley: University of California Press, 2010.

Annual Report of the American Museum of Natural History, vols. 30–35.

Brinkman, Paul. *The Second Jurassic Dinosaur Rush.* Chicago: University of Chicago Press, 2010.

Chapter Eleven
THE BONES OF THE KING

Gray, Christopher. "Streetscapes: The Dorilton; A Blowzy 1902 Broadway Belle." *New York Times,* September 30, 1990.

The Horseless Age: The Automobile Trade Magazine, February 19, 1902.

Browne, John. *Seven Elements That Changed the World*. New York: Open Road Media, 2014.

Annual Report of the American Museum of Natural History, 1903.

Dingus, Lowell, and Mark A. Norell. *Barnum Brown: The Man Who Discovered T. Rex*. Berkeley: University of California Press, 2010.

Brown, Barnum. Unpublished notes. American Museum of Natural History Vertebrate Paleontology Archives.

Lewis, Meriwether, and Clark, William C. *Original Journals of the Lewis and Clark Expedition, 1804–1806*. New York: Dodd, Mead, 1904.

Gedden, Stanley. *Big Bone Lick: The Cradle of American Paleontology*. Lexington, KY: University of Kentucky Press, 2021.

Dingus, Lowell. *Hell Creek, Montana: America's Key to the Prehistoric Past*. New York: St. Martin's, 2015.

Chapter Twelve
NEW BEGINNINGS

Osborn, Henry Fairfield. "Tyrannosaurus and Other Cretaceous Carnivorous Dinosaurs." *Bulletin of the American Museum of Natural History* 21 (1905).

Black, Riley. "Everything You Wanted to Know About Dinosaur Sex." *Smithsonian*, February 10, 2011.

Wellnhofer, Peter. "Archaeopteryx." *Scientific American* 262, no. 5 (May 1990).

Norman, Andrew. *Charles Darwin: Destroyer of Myths*. New York: Skyhorse, 2014.

Moore, Randy. *Dinosaurs by the Decades*. Santa Barbara, CA: Greenwood, 2014.

Rieppel, Lukas. "Bringing Dinosaurs Back to Life: Exhibiting Prehistory at the American Museum of Natural History." *Isis* 103, no. 3 (September 2012).

Brown, Barnum. Unpublished notes. American Museum of Natural History Vertebrate Paleontology Archives.

Brown, Frances R. *Let's Call Him Barnum*. New York: Vantage Press, 1987.

Sullivan, Jill A. *Popular Exhibitions, Science and Showmanship, 1840–1910*. London: Routledge, 2015.

"Sir Edwin Ray Lankester." *Nature* 159, 734 (May 31, 1947).

Chapter Thirteen
THE HARDEST WORK HE COULD FIND

"The Prize Fighter of Antiquity Discovered and Restored." *New York Times*, December 30, 1906.

Brown, Barnum. Unpublished notes. American Museum of Natural History Vertebrate Paleontology Archives.

Brown, Frances R. *Let's Call Him Barnum*. New York: Vantage Press, 1987.

Dingus, Lowell, and Mark A. Norell. *Barnum Brown: The Man Who Discovered T. Rex*. Berkeley: University of California Press, 2010.

Chapter Fourteen
A NEW WORLD

Brown, Barnum. Unpublished notes. American Museum of Natural History Vertebrate Paleontology Archives.

Brown, Frances R. *Let's Call Him Barnum*. New York: Vantage Press, 1987.

Dingus, Lowell, and Mark A. Norell. *Barnum Brown: The Man Who Discovered T. Rex*. Berkeley: University of California Press, 2010.

Prothero, Donald R. *The Story of the Dinosaurs in 25 Discoveries: Amazing Fossils and the People Who Found Them*. New York: Columbia University Press, 2019.

Sternberg, Charles Hazelius. *The Life of a Fossil Hunter*. New York: H. Holt, 1909.

Doyle, Arthur Conan. *The Lost World*. New York: A. L. Burt Co., 1912.

Purdy, Jedediah. "Environmentalism's Racist History." New Yorker .com, August 13, 2015.

Dickerson, Mary Cynthia (editor.) *Natural History: The Journal of the American Museum of Natural History. Volume XX.* New York City: The American Museum of Natural History, 1920.

Eugenics, Genetics and the Family: Scientific Papers of the Second International Congress of Eugenics, Held at the American Museum of Natural History, New York. September 22–28, 1921. Vol. 1

Canemaker, John. *Winsor McCay: His Life and Art.* Boca Raton, FL: CRC Press, 2018.

Chapter Fifteen
THE MONSTER UNVEILED

"Battle of Loos." National Army Museum. Accessed at nam.ac.uk/explore/battle-loos.

"It's the Red Sox Again and Again It's 2 to 1." *New York Times,* October 9, 1915.

Annual Report of the American Museum of Natural History, 1949.

Paul, Gregory S. "The Art of Charles R. Knight." *Scientific American* 274, no. 6 (June 1996).

Dingus, Lowell, and Mark A. Norell. *Barnum Brown: The Man Who Discovered T. Rex.* Berkeley: University of California Press, 2010.

Brown, Frances R. *Let's Call Him Barnum.* New York: Vantage Press, 1987.

Hall, Mordaunt. "The Screen; A Wonderful Farce." *New York Times,* February 9, 1925.

Hall, Mordaunt. "A Fantastic Film In Which A Monstrous Ape Uses Automobiles for Missiles and Climbs a Skyscraper." *New York Times,* March 3, 1933.

Chapter Sixteen
A SECOND CHANCE

"The Day War Came to the Polo Grounds." *Sports Illustrated,* October 24, 1966.

Jamieson, Wendell. "The Panic After Pearl Harbor: NYC on the Cusp of WW2." *New York Daily News,* August 14, 2017.

Shirley, Craig. *December, 1941: 31 Days That Changed America and Saved the World*. Nashville: Thomas Nelson, 2011.

Nicholas, Lynn H. *The Rape of Europe: The Fate of Europe's Treasures in the Third Reich and the Second World War*. New York: Knopf Doubleday, 2009.

Dingus, Lowell, and Mark A. Norell. *Barnum Brown: The Man Who Discovered T. Rex*. Berkeley: University of California Press, 2010.

Brown, Frances R. *Let's Call Him Barnum*. New York: Vantage Press, 1987.

"Birth Control Peril to Race, Says Osborn." *New York Times*, August 23, 1932.

Keith, Sir Arthur. "Whence Came the White Race?" *New York Times*, October 12, 1930.

Macdonald, William. "Mr. Grant's Plea for a Nordic, Protestant America." *New York Times*, November, 5, 1933.

Brown, Barnum. "A Luminous Spider." *Science* 62, no. 1599 (August 21, 1925).

Brown, Lilian. *I Married a Dinosaur*. Landisville, PA: Coachwhip, 2010 (reprint).

"Barnum Brown Dies at 89; Noted Collector of Fossils." *New York Times*, February 6, 1963.

Epilogue
THE MONSTER'S TRACKS

Small, Zachary. "A T. Rex Skeleton Arrives in Rockefeller Center Ahead of Auction." *New York Times*, September 16, 2020.

"The Life and Times of Stan, One of the Most Complete T. Rex Skeletons Ever Found." Christie's online catalog. Accessed at https://www.christies.com/features/The-life-of-STAN-a-T-rex-excavated-in-1992-10872-$2aspx.

Freedom du Lac, J. "The T. Rex that Got Away." *Washington Post*, April 5, 2014.

Browne, Malcolm W. "Tyrannosaurus Skeleton Is Sold to a Museum for $8.36 Million." *New York Times*, October 5, 1997.

Small, Zachary. "T. Rex Skeleton Brings $31.8 Million at Christie's Auction." *New York Times*, October 6, 2020.

Holson, Laura M. "He Listed a T. Rex Fossil on eBay for $2.95 Million. Scientists Weren't Thrilled." *New York Times*, April 17, 2019.

Holson, Laura M. "'Scotty' the T. Rex is the Heaviest Ever Found, Scientists Say." *New York Times*, March 28, 2019.

Chang, Kenneth. "How Many Tyrannosaurs Rexes Ever Lived on Earth? Here's A New Clue." *New York Times*, April 15, 2021.

Gorman, James. "Tyrannosaurus Rex: The Once and Future King." *New York Times*, March 4, 2019.

INDEX